"Structure is a science? Structure is more fundamental than processes? Than strategies? Than even people? Shared-services is not charge-backs? Every manager is an entrepreneur? Dotted lines/matrix/federated models — all wrong? Wow, this book blew away the fads and 'common wisdom' (which turns out to be unwise), and opened my eyes to a completely new, refreshing, powerful, and pragmatic understanding of organizations. A great read. Great work."

Chris S. Romano, CIO, Legal Industry

"I've been part of the senior leadership of a variety of organizations — start-ups to large, and medium- to high-growth companies — and I never looked at organizational design as a science. This book makes tremendous sense to me. It's profound, and at the same time, practical."

Ulrich Pilz, Board Director, multiple companies

"Other books I've read on structure present lots of theories, but no clear guidance. This is breakthrough thinking, and yet Dean makes it so practical and accessible that is seems like common sense."

Pat Beach, Director, Corporate IT, SRC Inc.

"This is the first book I've read on organizational structure that makes sense in the real world, hangs together, and comes with an instruction manual on how to implement change."

Fred Dewey, CEO and Serial Entrepreneur

"In this era of dramatic change fueled by technology, enterprises large and small will inevitably restructure to add new roles and functions, and to create new levels of cross-functional collaboration. Dean Meyer brings a set of principles and practices for organizational structure that every executive, from the CEO down, can use to maximize organizational effectiveness. Every leader with responsibility for structuring any part of the enterprise will profit from the lessons Meyer lays out here."

Richard Hunter, VP and Distinguished Analyst, Gartner

"This book gives me a transformational approach to redesigning my organization. It's based upon solid engineering and business practices. It tells me how to build really effective enterprise teamwork. And the suggested implementation process ensures success through the engagement of my staff in the design process. As a bonus, it's well written and easily understood, with frequent summaries of key points."

Karan Powell, President, American Public University

"Market leadership depends on innovative growth strategies, translated into deployable growth initiatives, and an organization that can execute. <u>Principle-based Organizational Structure</u> lays out how to build that organization."

Michael Treacy, best-selling author, <u>The Discipline of Market Leaders</u>

"I've applied these principles to my entire company, and it has been an engine for our growth."

Sergio Paiz, CEO, PDC

"As the CEO of a global leader in building automation, this book showed me how to better align regional organizations with the corporate strategy."

Lars vander Haegen, CEO, Belimo

"This book systematically transforms the mystique of organizational design into an open, rational engineering science. It explains how to mature an organization without losing its entrepreneurial spirit, a key ingredient of 'crossing the chasm.'"

Geoffrey Moore, best-selling author, <u>Crossing the Chasm</u>

"Our multinational chain of fashion and cosmetics retail stores is not a large company; but nonetheless, we had organizational issues. And I had difficulty sorting out individual performance problems from structural problems. Now, thanks to these Principles, my leaders have clear accountabilities; we understand how we team; and we can take on new brands and open new stores without having to make arbitrary assignments or rethink our structure."

Alejandro Arzu, President and CEO, Multinational Brand Management Group

"I've implemented these principles in organizations as small as 20 people (as well as much bigger ones), and it's always produced powerful insights, and practical and effective organization charts that drove real change."

Gary Rietz, CIO, Blommer Chocolate Company

"Frankly, after studying and applying these principles repeatedly, I can't imagine why other executives would depend on their intuitions, or repeat the mistakes others have made under the guise of 'best practices.' I encourage you to read this book before you make any decisions about organizational structure."

Preston T. Simons, CIO, Aurora Health Care and former CIO, Abbott Laboratories

Principle-based Organizational Structure

*a handbook to help you engineer
entrepreneurial thinking and teamwork
into organizations of any size*

by

N. Dean Meyer

PRINCIPLE-BASED ORGANIZATIONAL STRUCTURE:
a handbook to help you engineer entrepreneurial thinking and teamwork into organizations of any size

Meyer, N. Dean

Key words: organizational structure, organization charts, job design, organic organization, network organization, cellular organization, silo organization, matrix organization, decentralization, outsourcing, business process reengineering, dotted lines, Holacracy, transformation, leadership, empowerment, roles and responsibilities, teamwork, organizational effectiveness, organizational development, shared services, entrepreneurship, intrapreneurship, business within a business.

NDMA Publishing
3-B Kenosia Avenue
Danbury, CT 06810 USA
203-790-1100
ndma@ndma.com

Copyright 2017.

All rights are reserved. This document, or parts thereof, may not be reproduced in any form without the written permission of the publisher.

ISBN 1-892606-32-1

Printed in the United States of America.

*To the leaders I've worked with over the years
who helped me refine this science of structure
in the course of applying it to their organizations,
especially Preston Simons and Sergio Paiz.*

My thanks to the many leaders over the decades who helped me refine and test the principles in this book.

A special thanks to those who also gave me feedback on the manuscript itself, including:

Alejandro Arzu

Pat Beach

Mitch Betts

Fred Dewey

Robert Epstein

Richard Hunter

Harvey Koeppel

Geoffrey Moore

Sergio Paiz

Rainer Paul

Ulrich Pilz

Karan Powell

Susan Rho

Gary Rietz

Chris Romano

Preston Simons

Jeff Stovall

Michael Treacy

Lars vander Haegen

Richard Welke

CONTENTS

CASE STUDIES	ix
FIGURES	xi
FOREWORD: A CEO Who Has Applied the Principles of Structure	xiii
PART 1: Why You Should Read this Book	1
What you'll get from this book, including the importance of structure and why it's a science.	
1: The Importance of Structure	7
2: The Science of Structure	13
PART 2: Fundamental Principles	21
Seven fundamental Principles that apply to the design of organization charts.	
3: Principle 1: Golden Rule: Authority and Accountability Must Match	23
4: Principle 2: Specialization and Teamwork	32
5: Principle 3: Precise Domains	45
6: Principle 4: Basis for Substructure	60
7: Principle 5: Avoid Conflicts of Interests	67
8: Principle 6: Cluster by Professional Synergies	78
9: Principle 7: Business Within a Business	84

PART 3: The Building Blocks of Organization Charts   97

All the lines of business that exist within organizations, which become the elements of an organization chart.

10:	Overview of the Building Blocks	98
11:	Distinguishing Engineers and Service Providers	101
12:	Engineers	103
13:	Service Providers	109
14:	Coordinators	115
15:	Sales and Marketing	126
16:	Audit	136

PART 4: Applying the Principles, Seeing the Problems   141

Applying the Principles, you'll see how to look at any organization chart and anticipate its problems.

17:	Diagnosing an Organization Chart: The Rainbow Analysis	142

PART 5: Structures Designed to Fail   153

A series of case studies that warn you about mistakes others have made.

18:	Strategy as a Basis for Structure	154
19:	Pick a Core Competency, Outsource the Rest	158
20:	Managers as Client Liaisons	165
21:	Decentralization	171
22:	Dotted Lines and Matrix Structures	184
23:	Plan-Build-Run	189
24:	New Versus Old	193
25:	Quick Versus Slow (Bi-modal)	198
26:	The Pool	203
27:	Boundaryless/Network/Cellular/Organic Organization	206
28:	Leaderless Organization (Sociocracy and Holacracy)	213

PART 6:	**Design: How to Assemble the Building Blocks into an Organization Chart**	**223**
	How to assemble the Building Blocks into an organization chart tailored to your needs.	
29:	Clean Sheet Versus Tweaks	224
30:	Assumptions	227
31:	All the Lines of Business	235
32:	Clustering the Lines of Business	239
33:	From Clusters to Boxes	244
PART 7:	**Special Situations and Design Guidelines**	**255**
	Special situations and design guidelines, to help you with the design of your organization chart.	
34:	Self-managed Groups	256
35:	Shared People: Temporary Duty	259
36:	Remote Locations	262
37:	Project Management Office	267
38:	Compliance and Governance	270
PART 8:	**Workflows: You Can't Specialize If You Can't Team**	**279**
	How to build an explicit, but flexible, process of cross-boundary teamwork.	
39:	What *Not* to Do to Improve Teamwork	281
40:	High-performance Teamwork	287
41:	The Limits of Traditional Approaches to Teamwork	290
42:	The Teamwork Meta-process: Concepts	300
43:	The Teamwork Meta-process: Mechanics (Walk-throughs)	308

PART 9: The Process of Restructuring   **315**
Implementation, with a step-by-step project plan for restructuring and for consolidations.

44: Change Management   317
45: The Restructuring Process, Step by Step   325
46: Establishing the Sales Function in an Internal Service Provider   336
47: Consolidation of Shared Services and Acquisition Integration   342

PART 10: Leadership Questions   **351**
The benefits that justify the costs; where structure fits in your transformation strategy; and how to get started.

48: Should I Do This? Benefits That Justify the Costs   352
49: When? Place Within a Transformation Strategy   363
50: What Should I Do Next? Ideas to Action   371

APPENDIXES   **375**
Appendix 1: Terminology   376
Appendix 2: Supervisory Duties   382
Appendix 3: Theoretical Underpinnings   385

ENDNOTES (References)   **393**

INDEX   **405**

Other Books by N. Dean Meyer   421

About the Author   422

CASE STUDIES

A CEO Who Has Applied the Principles of Structure	xiii
Combining Invention and Operations: Airline	9
Combining Sales and Coordination: Higher Education	16
Process Owners	23
The CFO Who Mandates Others' Budgets	25
Customer Service Held Accountable for Resolving Incidents	25
Mandating "How" Instead of "What"	27
Customer-centric Structure	32
Vague Domains — IT Infrastructure versus Enterprise Architecture	45
Gaps — Layers of Engineering	48
The Evils of Roles and Responsibilities	52
Structure by Clients' Business Processes	60
The Governance and Client Liaison Group	67
Combining Invention and Operations: IT	69
Combining Invention and Operations: Corporate Acquisitions	71
Benefits of Separating Purpose-specific Solutions from Components	73

Process-centric Groups	78
Safety Group that's Accountable for Safety	84
Managers as Client Liaisons	165
Decentralization: Manufacturing Plants	173
Decentralization: Design Engineering	173
Decentralization: IT	174
Plan-Build-Run	189
New Versus Old	193
Quick Versus Slow (Bi-modal)	198
The Pool	203
Holacracy: Disempowerment of Management	218
Good Reasons for the Wrong Basis for Substructure: Sales by Product Type	251
The PMO that Manages Projects	267
Chief Compliance Officer Accountable for Compliance	270
What *Not* to Do to Improve Teamwork	281
Silo Organization: Manufacturing Plants and Scheduling	291
Transformational Benefits	358

FIGURES

Figure 1: Principles of Structure   22

Figure 2: The T-shaped Specialist   37

Figure 3: The Building Blocks of Structure   99

Figure 4: Applications versus Base Engineers   105

Figure 5: Competencies of Engineers   107

Figure 6: Biases of Engineers   108

Figure 7: Three Types of People-based Service Providers   112

Figure 8: Competencies of Service Providers   113

Figure 9: Biases of Service Providers   114

Figure 10: Competencies of Coordinators   124

Figure 11: Biases of Coordinators   125

Figure 12: Three Types of Sales and Marketing   131

Figure 13: Competencies of Sales and Marketing   134

Figure 14: Biases of Sales and Marketing   135

Figure 15: Competencies of Audit   139

Figure 16: Biases of Audit   139

Figure 17: Four Questions of the Rainbow Analysis   145

Figure 18: Consequences of Gaps   146

Figure 19: Consequences of Rainbows	147
Figure 20: Consequences of Scattered Campuses	148
Figure 21: Bases for Substructure	149
Figure 22: Consequences of Inappropriate Substructure	150
Figure 23: Two Types of Dotted Lines	184
Figure 24: Relationship with Shared Services (Model A/B)	237
Figure 25: Alternatives for Remote Locations	265
Figure 26: Format of a Walk-through	309
Figure 27: Overview of the Restructuring Process	327
Figure 28: Five Organizational Systems	363
Figure 29: Two Components of Organizational Structure	378
Figure 30: "Customer" Versus "Client"	381

~ FOREWORD ~

A CEO Who Has Applied the Principles of Structure

by Sergio Paiz, CEO, PDC
May, 2016

My brother Salvador and I took over the family business after my father passed away in a tragic airplane accident. I was 25, a recent engineering graduate, and new to the world of business.

While my inexperience was a source of immense stress, it may have been an advantage as well. I looked at the world through fresh eyes.

What I saw was a loosely related set of companies, with around 850 professional staff and total revenues of approximately US$130 million. And most had serious problems.

On the positive side, these companies had tremendous potential for synergies and growth. I saw so many untapped opportunities.

The Problem

Sales had grown threefold over eight years. With a proliferation of new lines of business, the number of departments burgeoned from 38 to 67. Then we hit the wall, and growth stopped. We had outgrown our physical and IT infrastructure, and systems were breaking down. Our business model was not scalable.

I questioned why we needed three different call centers, for example. Business-unit leaders insisted that, if they're to be held accountable for their results, they needed all the support functions reporting to them.

Some actually threatened to resign if I interfered with their fiefdoms. In their minds, a large staff gave them power. It didn't help that our job-grading system rewarded those with more headcount.

Meanwhile, every company had different ERP systems, accounting processes, compensation systems, etc. It was tough to get a consolidated view of the firm.

Furthermore, we struggled with a lack of sufficient talent to go around. With all this decentralization, we couldn't attract strong leaders or needed specialists to any of these functions.

Perhaps most telling, the company was operating at a heavy loss. I had banks calling me, concerned about our high levels of debt and deteriorating financial results.

Organizational Systems Approach

It seemed I was confronted with a new crisis each day. As I tackled one problem, ten more popped up. It was like patching a leaking boat, and I was the captain sinking with it!

I couldn't go on solving problems one at a time. After all, I only had so many hours in a day. I knew I needed to focus my attention on *root causes,* and design an *organizational system* that would solve problems on an ongoing basis, with or without me. I thought about the situation, and came to two conclusions:

One, of all the aspects of our organizational system, I needed to address structure first.

Two, I knew there was no way I could be alone in these problems. Thousands of other companies must experience these growing pains. I couldn't afford organizational design mistakes or disruptive experiments; that's way too expensive and time consuming. Trial-and-error was not an option. I needed help.

Discovering the Principles of Structure

As is customary whenever I have a problem to solve, I went in search of a "how to" book, this time on organizational design. I read several, most HR oriented, where structure was depicted as an art rather than a science. They essentially said that structure is the product of the style and preference of the CEO.

It was frustrating. I found no real answers in that view. Surely there must be a scientific approach to structure.

Finally, I came across the work of Dean Meyer. I read his book on structure. In fact, I read it twice, highlighted it, and summarized it. Then I read other books by Dean, for example, one on decentralization and one on leading transformations. This was the systemic approach I'd been looking for.

Our Restructuring Process

I arranged a telephone call with Dean, and he explained how we could design and implement a new structure in an open, participative process that engaged our leadership team. First by phone and then in person, Dean and I crafted a detailed project plan, as well as a communication plan (which was equally important).

Then I announced the plan to our leadership team. My message was brief:

"We need to restructure our company. My promise to you is that no one will lose employment because of organizational changes. However, all of you will lose your titles as we work together on a 'clean sheet of paper' design. Please throw your business cards in the trashcan as you leave this room. I will rehire you into the new positions that we'll design together."

You could hear a pin drop. People were stunned. Some were furious. I had just destroyed their fiefdoms.

But with Dean's facilitation, we got them working on the design of the new organization. Dean taught them the Principles and the Building Blocks described in this book. Then, we analyzed which lines of business (functions) were present in each department. We even color-coded them by Building Block (the Rainbow Analysis described in Chapter 17), and it was quite a colorful chart!

The gravity of the problem quickly became evident to everyone:

- **Fragmentation:** Functions were scattered. There were no real centers of excellence. The color coding made the extent of the problem very graphical; and people began to understand the waste, lack of economies of scale, and lost synergies.

- **Gaps:** There were many critical functions that everybody thought somebody else was doing, but in fact no one was really accountable for them.

- **Rainbows:** The color-coding revealed a lot of "rainbows." Everybody was doing everything. With multiple functions under each leader, nobody could focus and specialize to the point of being particularly good at anything.

We put all the lines of business on post-it notes, and sorted them into the five Building Blocks on posters around the meeting room. Then, we jointly designed the new structure, sorting those post-its into stacks that would represent the departments under me.

I was afraid this would be a political free-for-all, with everybody fighting to rebuild their fiefdoms. But in fact, the Principles and Building Blocks, along with Dean's facilitation, led us to a fact-based discussion of what would be best for the company.

The New Structure

To their credit, most participating leaders understood the benefits of becoming one integrated organization. And Dean explained how everybody could be an empowered entrepreneur running a small business within a business, and how teamwork could give everybody access to the right specialists anywhere in the company, without having to have them reporting to you.

So the leadership team opted for a fully consolidated structure. We established product managers (profit centers) for each of the product lines — a list that included food and household products, lubricants, and real-estate financial services.

Everything else became an internal shared service. Yes, everything! That includes sales, marketing, manufacturing, logistics, customer service, business planning, as well as all the traditional support services like IT, HR, and Finance. All were there to serve the product managers (and each other).

We opted for the ideal — the maximum in synergies and economies of scale. And every box on the organization chart was designed as a business within a business — an entrepreneurship chartered to serve customers inside the company or outside.

The new structure was simpler, cleaner, and easier for everyone to understand.

And then, we started investing in teamwork, understanding how all these departments provide their services to companywide teams (the process described in Part 8 of this book).

The Hardest Part

There certainly was some resistance, with a few leaders still trying to build fiefdoms. And there were leaders who just weren't able to understand the new operating model; it wasn't the way they'd always done things, or they didn't have the leadership skills to succeed in that empowered, accountable environment.

A challenge for me was resisting the temptation to compromise the structure to accommodate people. In fact, the hardest part of my job was letting go the people who needed to leave, and recruiting the right talent. This has taken a few years. It was personally painful, and drained huge amounts of my energy. But it was the right thing to do — really the only thing to do.

Impacts on Culture

This new structure hinges on a business-within-a-business model. Every group is defined as a business that "sells" products and services to customers, be they inside or outside the company. (But we don't cross-charge for internal services.)

This requires that managers behave as entrepreneurs. I expect them to deliver high levels of customer service at competitive rates (costs), and to manage their businesses as if they were their own.

This kind of thinking is really starting to take hold. And the contrast is remarkable.

Before the restructuring, business units were insular, and the culture was hierarchical. Now, it's open, empowered, and collaborative.

Before the restructuring, corporate functions thought their job was to control the business units, not serve them. Now, every function is customer focused and service oriented.

Before the restructuring, all except the most senior executives believed their job was to manage resources and processes. Now, everyone is accountable for delivering results.

As we empowered our internal entrepreneurs, motivation went through the roof. People couldn't believe they had the authority to run their businesses their way! Of course, we were very careful to make sure that accountabilities matched managers' authorities.

Our entrepreneurial culture unleashed everyone's creativity. Here's one great example: A manager of an internal support function told me that his team wanted to grow their business by selling their services to external customers. We analyzed this and decided it would be a good idea. It would create economies of scale that would reduce internal costs; and it would improve the quality of internal services, while contributing additional profits.

Then, other areas began to sell services outside the organization, including IT, Logistics, and Merchandising Services. Entrepreneurship was contagious!

Our new entrepreneurial spirit applies within the company as well. For example, our Health, Safety and Environment (HSE) group was doing great work in our manufacturing sites. When these bright people were empowered to run HSE as a business, they sold their services to other departments like Logistics and Administration.

Our Operations Research team extended their services beyond truck-route optimization. Now they're helping the Warehouse optimize layouts, and Sales plan its territories and routes.

Also, in the past, staff had little regard for costs. Budgets were padded, and cost control was left to our Finance group (and ultimately to me).

Now, managers have a clear understanding of what businesses they're in, and who their competition is (outsourcing). So everybody is more cost conscious.

While it's not a comprehensive treatment of culture, structure has had a tremendous positive effect on the way we think and work.

I've seen some seasoned managers struggle with these changes in our culture. I think that's good. It's developing real leadership skills. Another nice benefit is that some young talent is thriving in this environment; these are our future senior leaders.

Business Results

At the beginning of our transformation, we lost sales and EBITDA. This was in part due to our shedding companies where we couldn't buy out other shareholders, and divesting or closing several less-profitable lines of business.

But of course the restructuring had a cost, too. It took a lot of our leaders' time and attention, which may have taken our eyes off business challenges to some degree. Plus, we've had to invest in bringing in some qualified senior leaders that we'd been missing in the past structure, for example, over the consolidated sales, engineering, and manufacturing groups.

But after that initial stage, the new structure has been an engine for our growth.

Part of that growth came from improved performance in every function. People are better focused and more specialized, and individual accountabilities are much clearer. This filtered out the low performers, and enhanced everybody else's performance.

Career paths were a problem when little pockets of a given profession were scattered among our various business units. Now, we're able to attract excellent talent into well-focused jobs, with better career paths in each profession. Staff can focus on different sub-specialties. And economies of scale have enabled more training, better tools, and improved infrastructure.

Also, we eliminated redundancies and drastically reduced internal friction. This saved money, and improved staff's job satisfaction. We now have a very positive, cohesive team.

Beyond that, we found business synergies which otherwise couldn't have happened. For example, we achieved tremendous efficiencies in Transportation when we integrated the traffic from all our business units into a single delivery service.

In addition to cost savings, this structure increased revenues. For example, our lubricants business used to be run as a separate company, selling lubricants to industrial customers. Now that all product lines share sales and distribution services, we also sell oil to consumer outlets such as convenience stores within gas stations. This has resulted in growth for the brand.

Our structure is highly scalable. For example, we acquired a bleach company. Everything integrated perfectly into our well-designed structure, and everybody was clear on what to do. Bleach became a new product line within our existing product-management group. Its manufacturing plants fit in nicely under our existing operations group. Bleach is distributed through our existing warehouses and transportation services. And of course it's sold through our existing sales forces.

And when we won the distributorship of lubricants in a new country — a huge increase in revenues for us — we were able to clearly define accountabilities for its success in all our functions.

Of course, the real proof is in the numbers. In the past three years (since the restructuring), we're back on track with growth in sales (averaging 22 percent per year). And after a decade of losses, we are now profitable. In fact, EBITDA has been growing at a sustainable 44 percent annually.

I honestly don't think we could have attained nearly these results without our new structure. And this is just the beginning.

I am certain this scientific approach to organizational structure has profoundly transformed our company. It's given us the right framework to continue building a profitable and scalable company for many years to come. I'd have to say that a principle-based structure has been one of the best investments I've made.

Principle-based Organizational Structure

*a handbook to help you engineer
entrepreneurial thinking and teamwork
into organizations of any size*

~ PART 1 ~
Why You Should Read this Book

*When you were made a leader, you weren't given a crown;
you were given the responsibility to bring out the best in others.*

Jack Welch, CEO, GE

Few executives succeed because of their individual efforts alone. As an executive, you're driving a "machine" — the organization under you. That machine can either get you to your vision, or it's the albatross that frustrates your ambitions.

Sergio Paiz once observed, *"Most managers focus on operating a poorly designed machine, and struggle with it, rather than stepping back and redesigning the machine."*

Consider this leader who knew better, described in a research report summarized in the *Harvard Business Review*. [1]

Google's astonishing success... depended in large part on its ability to... scale up its infrastructure at an unprecedented pace. Bill Coughran, as a senior vice president of engineering, led the group from 2003 to 2011 [that]... built Google's "engine room"....

Given Google's ferocious appetite for growth, Coughran knew that GFS [Google File System] — once a groundbreaking innovation —

would have to be replaced within a couple of years.... One might expect that he would first focus on developing a technical solution for Google's storage problems and then lead his group through implementation.

But that's not how Coughran proceeded. To him, there was a bigger problem, a perennial challenge that many leaders inevitably come to contemplate: How do I build an organization capable of innovating continually over time? Coughran knew that the role of a leader of innovation is not to set a vision and motivate others to follow it. It's to create a community that is willing and able to generate new ideas.

Whether you lead a company or a department within a company, one of the most important things you can do as a leader is build a machine that's capable of taking you where you want to go.

This is why great leaders pursue **three strategic vectors** in parallel: **market** (alignment with customers), **technical** (product/service capabilities), and **organizational** (the machine). [2] This book is about your organizational strategy.

Key to the design of an organizational machine is its **structure — the organization chart that defines everybody's domains, and the processes that combine those specialists on teams**. Sure, there are other elements of organizational design, such as culture, resource-governance processes, methods, and metrics (discussed in Chapter 49). But of all these, structure is the most powerful.

Here are a few of the many **benefits** of a healthy structure:

- Individual accountabilities for results are clear; so people deliver results.

- Staff are focused on excellence in a single profession, so they perform better.

- Staff are customer focused; relationships with clients improve; the organization is well aligned with clients' needs and strategies; and synergies across clients' businesses are discovered.

- Teams form spontaneously, and work well together in flexible, but well-defined, processes that are tailored to the needs of specific projects and services.

- The organization delivers its commitments reliably, more quickly, and with lower risk.

- Costs are competitive as redundancies are eliminated and internal entrepreneurs are accountable for offering best value.

- Product quality goes up; and the product line is better integrated.

- Staff are creative and entrepreneurial. The pace of innovation in every discipline improves.

- The organization is a great place to work because staff are empowered, conflicts of interests are eliminated, and jobs don't expect the impossible.

- The structure is lasting; it's not designed around incumbent personalities; it defines accountabilities for all domains (current and future); and it automatically evolves as strategies and technologies change.

In short, a well-designed organizational structure contributes directly to shareholder value (i.e., the mission of an organization).

On the other hand, a poorly designed structure causes far-reaching problems. You may recognize some of these **symptoms**:

- Unclear individual accountabilities for results; staff who are task or process (rather than results) focused; confusion about who does what; redundant efforts; territorial disputes; internal competition

- Poor performance because people are going too many ways at once; jobs which are too big for most people to succeed; a need for all "A players"

- Lack of customer focus; weak or strained relationships with clients; initiatives which are product or technology (rather than business) driven

- Difficulties with cross-boundary teamwork; slow response to new challenges; an organization of independent "silos"

- Slow or unreliable delivery

- High costs; a lack of concern for frugality; empire building

- Poor quality; a poorly integrated product line

- Lack of entrepreneurial spirit; staff aren't creative, and don't take initiatives to improve their functions

- Lagging in innovation; lack of accountability for planning and creating the future

- Low morale and motivation; disempowerment; dead-end jobs; poor morale; cynicism; stress

- Repeated restructurings, each fixing some problems and creating others

An unhealthy structure is not a "supplier of choice" to its customers (internal or external), nor is it an "employer of choice" to its staff. It's a stressful place to work. And it's a tough organization to lead; it takes lots of your attention to keep it running right, leaving you little time for strategic thinking.

This book will help you pinpoint any problems in your current structure, and design the ideal structure for your organization.

So what's in this book? Here's the "**elevator pitch**":

- **Science:** Contrary to many people's beliefs, organizational structure is an engineering science. (It's not a matter of personalities, politics, and personal style.) And while there's no one right structure that fits every organization, there absolutely is a pragmatic way to engineer the best organizational structure for you.

- **Principles:** Seven firm principles are the foundation of the science of structure. For example: To succeed, people's authorities must match their accountabilities. Specialization leads to better performance. And the way you divide a function into groups tells people what they're supposed to be good at.

- **Business within a business:** The best organizations induce staff to think and behave as empowered entrepreneurs, running small businesses within a business. They serve customers (external and internal) with defined products and services; as such, they're aligned with customers' needs.

  There are five types of businesses present in any organization — the "Building Blocks" of structure. This framework is a language for describing organization charts. It helps to ensure that the entire mission of your organization is covered, without gaps or overlaps.

- **Teamwork:** Effective teamwork is essential. Without it, groups revert to independent "silos" no matter what the organization chart says.

  Effective teamwork is not a matter of team-building to improve personal relationships and trust. And except for simple assembly lines, it's not a matter of process engineering. Great cross-boundary teamwork is based on a "meta-process" that flexibly combines just the right talent on each team, and customizes processes to the unique needs of each project or service.

- **Participative change process:** The process of change is critical. People must understand the new structure, believe in it, and be committed to making it work. The best way to achieve that is an open, participative implementation process.

  A well-proven, step-by-step process has evolved over more than three decades of study and dozens of implementations. It can be used to restructure an existing organization, or to merge organizations in a shared-services consolidation or acquisition integration process.

This first Part discusses why reading this book is worth your time.

Chapter 1 explores the power of structure, and why just hiring great people or defining processes isn't enough.

Chapter 2 explains why designing your organization's structure is a challenge worthy of study and careful thought, and gives you an overview of the *science* of structure.

Chapter 1:
The Importance of Structure

It wouldn't be surprising if you felt cynical about the importance of organizational structure.

Sadly, most of us have experienced restructurings that did little more than "rearrange the deck chairs on the Titanic." Too often, a reorganization is just a means of accommodating the careers of the top few, or a way for a new executive to "stir the pot." But for most staff, work goes on as it always has — perhaps with a new boss to put up with, perhaps with new political tensions replacing old ones, but with little real change in what people do or how well the organization performs.

Were these ill-fated restructurings ineffective because organizational structure doesn't matter all that much? Or was it because these structures were poorly designed or badly implemented?

Before we dig into the science of structure (Chapter 2), let's address the question: Is structure really all that important?

Aren't Good People Enough?

Some argue that working relationships are formed *despite* the structure, creating an "informal organization" that really determines how things get done. They conclude that success depends simply on hiring and developing great people who work hard, work well together, and "do what's right" for the organization.

This line of thinking leads them to design organizational structures around the unique talents or career needs of key executives.

There are at least three reasons why this is a bad idea.

First, you have to restructure the organization whenever anyone changes jobs. Otherwise, successive managers, each with their own unique talents, will find those carefully tailored jobs impossible to fill.

Each of these restructurings is disruptive and expensive. And since repeated reorganizations do little to improve the organization's effectiveness, they reinforce the belief that structure doesn't matter — a self-fulfilling prophecy.

Second, organizations which depend only on great people require leadership teams that work so well together that structure doesn't matter. It requires that people look past their own job descriptions and personal interests for the good of the organization.

This altruism isn't natural; and it certainly isn't a reliable process. These organizations are fragile. They require a charismatic leader, and a tightly knit and dedicated leadership team. They generally fall apart when the executive who created the team or any of its key members leave (or just get tired).

Third, an organization that depends on everyone being "great" is impractical. Sure, great people are, well, *great*. But an organization which requires that everyone is above average is difficult to staff (because you can't always afford the best, or they may not be available), and limited in its growth potential (because you can only hire so many super-achievers).

Chapter 1: The Importance of Structure

It's true that great people can overcome any structural problems. Well, let me be more specific....

> Superior people who work well with their peers,
> > who are motivated by an inspiring leader,
> > > and who are willing to work long hours
> > > > and to set aside their own best interests (and those of their staff) to do what's best for the organization,
> > > > > can overcome most structural dysfunctions.

But if the structure is poorly designed, an inordinate amount of leaders' time, energy, and goodwill is spent resolving confusion and friction rather than doing real work. **Wouldn't you rather your great people focus their precious time and energies on achieving great things, than spend it dealing with self-imposed obstacles** which can (and should) be fixed?

Or worse, sometimes people just can't overcome the forces of structure. Here's a sad <u>case</u> example. [3]

A large airline divided its IT department into two computer centers: business systems, and the reservations system. Leading each, an "operations director" was tasked with keeping those critical systems running efficiently, safely, and reliably.

Reporting to each operations director was a systems-engineering manager, as well as a data-center manager.

In the business systems group, executives became very frustrated because IT retained an expensive old mainframe. The systems-engineering manager had failed to innovate, so they fired him.

Two years later, they fired his successor for the same reason.

And two years after that, still another systems-engineering manager was fired.

These managers were all bright, experienced people who were good at their professions. The problem wasn't their lack of ability.

Put yourself in the shoes of that systems-engineering manager. Your boss, the operations director, is paid to <u>keep things running efficiently, safely, and reliably</u>.

Would you dare to suggest a major innovation? Of course not! Your boss is rewarded for stability. And innovation inevitably disrupts operational stability. You were set up to fail.

No matter how good people may be, structure is a powerful force. If structure is causing people to do the wrong things, then "better" people will just do the wrong things faster.

Where you stand depends on where you sit.

Anonymous

Great people and a team spirit are, of course, wonderful. But they alone don't make an organization perform well. The truth is, **great people in a bad structure will fail,** or at least not perform to their potential. On the other hand, a normal bell-curve of people in a healthy structure can succeed.

Great organizations are, by design, those where average people can succeed, and super-achievers can super-succeed.

Aren't Good Processes Enough?

Others who doubt the importance of organizational structure believe that carefully engineered business processes are the key to success. They argue that as long as everybody does their part in well-designed processes, an organization will perform well.

This, too, is problematic for at least three reasons:

First, the effectiveness of processes *depends* on the organizational structure. If nobody specializes in a given discipline, processes which depend on that competency will fail. It's up to the organization chart to provide groups dedicated to every needed discipline.

Second, structure can override processes. Sometimes, processes ask people to do things at odds with their job descriptions. When staff are caught between conflicting forces, they'll optimize their performance appraisals, even if some processes fail. (It's always easy to blame others involved in the process for the failure!)

Third, predefined processes only make sense when work is highly structured and routine. But most organizations require much more than people performing routine tasks on a predefined assembly-line. In functions where projects have unique requirements, and where flexibility and innovation are critical components of success, simply giving everybody a clearly defined role in a carefully engineered process makes the organization more rigid and less creative.

The truth is, **great people working with great processes can't perform well unless the structure they live in is well designed.**

Structure Drives Performance

While good people and processes are important, structure is one of the most powerful drivers of an organization's performance — its efficiency, effectiveness, agility, quality, creativity, innovation, competitiveness, and customer satisfaction.

It's also a key element of staff's competence, job satisfaction, motivation, commitment, happiness, and loyalty.

Structure is the foundation on which good people and processes thrive.

SYNOPSIS

» Great people in a bad structure will fail, or at least not perform to their potential.

» Great processes don't work unless the structure is well designed.

» Structure is one of the most powerful drivers of an organization's performance.

Chapter 2:
The Science of Structure

Perhaps you agree that organizational structure is important. But does it take *study*?

As an executive, you've spent most of your adult life working within organizations. If you're lucky, you've seen a rare few that worked well. But in all likelihood, you've seen many that haven't.

With all that experience, perhaps you feel comfortable designing the structure of your own organization. After all, how hard can it be to draw some boxes and shuffle some names around!?

I know you've heard so many (often conflicting) theories, fads, and opinions about organizational structure over the years. And there certainly isn't one off-the-shelf organization chart that's right for everybody. Some academics even say that structure is contingent on your industry, style, or other factors. [4]

But despite all that noise, please understand this: There is a *science* of organizational structure.

---

science n. The observation, identification, description, experimental investigation, and theoretical explanation of phenomena.

American Heritage Dictionary

Organizations are systems, [5] and present an engineering challenge. And engineering certainly fits the definition of a science (an applied science, not a basic science, but with firm principles and constructs nonetheless).

> *"Engineers operate at the interface between science and society."*
> Gordon Stanley Brown, Dean of Engineering, MIT

The science of structure documented in this book evolved from real-world empirical observations that revealed patterns, generalized to principles and frameworks which were tested over decades of actual implementation experiences.

The principles are now so clear that you can **look at any organization chart and know who's fighting with whom, who's not making objectives, and who has ulcers!**

The Importance of Knowing That Science

When executives don't study the science of structure, reorganizations are guided by:

- Incumbents' personalities and careers; or attempts to work around managers who don't do essential aspects of their jobs
- Overly simplistic models (like "build versus run") or industry fads (like the matrix)
- Client pressures (such as groups dedicated to them)
- Today's work, priorities, and strategies
- Intuition

And when an organization's structure evolves through a series of incremental changes made by a number of different executives, each with their own philosophies and exigencies, it resembles a patchwork quilt of mismatched pieces.

Chapter 2: The Science of Structure

There are at least four reasons why executives should study the science of structure:

Organization du jour: Without guiding principles, each restructuring solves some problems while creating others. The results are generally far from satisfactory. What happens then? Another reorganization!

Some leaders restructure whenever there's a shift in business strategies, new technologies, new business initiatives, or when they hire a new leader. And whenever anything goes wrong, some executives restructure their organization again.

This "organization *du jour*" causes a never-ending, painful, and costly series of disruptions. Staff must endure an endless stream of unsettling changes. But in many cases, they see little real difference in the way things work. There are costs, but few benefits.

On the other hand, the science of structure helps you engineer organization charts systematically, making deliberate decisions about key trade-offs — just as an engineer does in any field of study.

The result: a structure that performs well at a multitude of challenges, and doesn't cause unintended consequences. And even in the long term, repeated reorganizations shouldn't be required. A principle-based structure dynamically adapts to changes in the world around it. It's a lasting investment.

Easier to explain: Another reason to study the science of structure is that it makes it easier for you to explain how things are supposed to work.

And it's easier for staff to understand it, so they're more likely to make it work as intended.

Also, with clear principles, an executive can invite participation from his/her leadership team in the design of a new structure without fear of endless haggling and parochial politics. Participation takes advantage of people's in-depth knowledge of the work of the organization. It builds their deep understanding of how the new structure should work. And participation engenders tremendous commitment to making the new structure a success.

Don't blame people: I'd argue that leaders have a moral obligation to study the science of structure, since it's critical to know the difference between a poorly performing person and a good person in a poorly designed job. Here's a case example:

In a growing accredited university, Sondra was hired to enhance corporate "outreach." She was given two goals:

1) Encourage corporations to utilize the University as part of their professional development programs (a sales function).

2) Build a "School of Continuing Studies" to deliver continuing education to these non-traditional students.

Sondra was good at sales. She pulled together existing corporate-outreach staff, and gave them the sales training they needed. She developed sales objectives, and a dashboard to monitor their performance. The result: She increased revenues by nearly $4 million in just one year!

But Sondra failed at her second objective.

The President had hoped she would sell the University's current course offerings to new audiences, perhaps repackaging them into

Chapter 2: The Science of Structure 17

smaller programs for badges or certificates, or defining new paths to existing degree offerings. To do this, Sondra would have to bring opportunities to the rest of the University (who already knew how to put together programs and deliver education) and coordinate enterprisewide initiatives.

Instead, Sondra came up with her own plan, as if she were setting up a parallel school. Her plan was inconsistent with the University's processes, and wasn't feasible.

Coordinating enterprisewide initiatives is a challenging job in its own right, requiring competencies quite different from sales. Sondra's failure was not because she was a poor performer, as her sales success proved. It was because the structure gave her two very different jobs, and no one could be expected to excel at both.

Fortunately, the President studied the science of structure, and saw the real problem before firing a great sales leader.

It's both futile and cruel to blame people for poor performance when structure is the root cause of problems. To know the difference, you need to understand the science of structure.

New levels of performance: On the positive side, a principle-based structure can perform far better. A great leader doesn't settle for less. Listen to the advice of Preston Simons, CIO at Aurora Health Care and former CIO at Abbott Laboratories:

As a CIO, I've led organizations that ranged from hundreds of employees regionally to thousands globally. And in each case, I've implemented these principles of structure.

This powerful, yet pragmatic, approach consistently leads to improved role clarity; better teamwork; customer focus; entrepreneurship; improved performance; and commitment.

I use the principles as a forensic tool to give me insights about the structure I've inherited. And they provide a basis for a participative change process that allows my leadership team to contribute their knowledge and build an organization that they truly understand and support.

Frankly, after studying and applying these principles repeatedly, I can't imagine why other executives would depend on their intuitions, or repeat the mistakes others have made under the guise of "best practices." I encourage you to read this book before you make any decisions about organizational structure.

What's to Come

This book explains the science of structure in practical, straightforward terms.

You can use it to analyze the pros and cons of your current organization chart, and see where your structure is getting in people's way.

You can use it to diagnose the faults in a proposed new structure.

On the positive side, this book gives you the tools to design and implement a high-performance organizational structure that's right for you. It provides a common language and set of principles as a foundation for objective, fact-based discussions with your leadership team. And there are case studies of structures that have failed throughout the book, especially in Part 5, to help you avoid the mistakes others have made.

Finally, it describes a restructuring process that's thorough, participative, and proven.

In all, this book can help you significantly improve performance; recruit and retain the best talent; adapt to changes in your mission or your environment; prepare for growth; and integrate mergers, acquisitions, and consolidations.

SYNOPSIS

» Organizational structure is an engineering science, with firm principles and constructs.

» Understanding this science helps you see the difference between poor performers and good people in poorly designed jobs.

» The science of structure can help you design the optimal structure for your organization, avoiding costly mistakes and precluding repeated disruptive restructurings.

» It can help you significantly improve performance; recruit and retain the best talent; adapt to changes in your mission or your environment; prepare for growth; and integrate mergers, acquisitions, and consolidations.

~ PART 2 ~
Fundamental Principles

> *As to methods there may be a million and then some, but principles are few.*
> *The man who grasps principles can successfully select his own methods.*
> *The man who tries methods, ignoring principles, is sure to have trouble.*
>
> Ralph Waldo Emerson

This book treats organizational structure as a matter of engineering a system. As in any engineering science, there are sound **principles** and clearly defined **components** to guide you. This Part focuses on the principles; Part 3 defines the components.

Then, Part 4 puts those two frameworks together into a diagnostic and design method; Part 5 exercises the science of structure by analyzing a series of case studies of structures designed to fail; and Parts 6 and 7 describe the design process.

So let's start with the principles.... Seven Principles provide the foundation for the science of structure. They're listed in Figure 1.

These seven Principles are described in the seven Chapters of this Part.

Figure 1: Principles of Structure

Principle 1: Golden Rule: Authority and accountability must match.

Principle 2: Specialization and Teamwork: You can only be world-class at one thing at a time; but you can't specialize if you can't team.

Principle 3: Precise Domains: Define clear boundaries with no overlaps or gaps.

Principle 4: Basis for Substructure: Divide a function into groups based on what it's supposed to be good at.

Principle 5: Avoid Conflicts of Interests: Don't expect people to go in two opposing directions.

Principle 6: Cluster by Professional Synergies: Cluster groups under a common boss based on similar professions.

Principle 7: Business Within a Business: Every manager is an entrepreneur whose job is to satisfy customers (internal and external) with products and services.

The Principles are illustrated by case studies which introduce the issues. For examples of departments within companies, I generally use IT organizations. IT is the largest and most complex of the internal support functions, and offers a rich microcosm of companies as a whole. **Regardless of whether a case study describes your industry or function, I hope you'll see the applicability of the concepts to your organization.**

Everything's got a moral, if you can only find it.
Lewis Carroll (Charles Lutwidge Dodgson)

Chapter 3:
Principle 1: Golden Rule:
Authority and Accountability Must Match

The first Principle is by far the most important — so important that I call it the "Golden Rule" of organizational design. It's absolutely essential to the success of every organization.

Before I state it, here's a case study that illustrates the converse.

Case Study: Process Owners

At the suggestion of process-improvement consultants, a CIO appointed a few "process owners." Each was assigned a process that engaged people from various parts of the organization in producing a specific service.

These process <u>owners</u> had authority over those processes. Sure, they were collaborative and involved stakeholders in designing and implementing new processes. But they were not process <u>facilitators</u> who served others by bringing teams to consensus on how they'll work together. Their job was to implement processes, and they had the final say.

However, these process owners didn't have matching accountability for the effectiveness of those processes — that is, they weren't always the ones accountable for delivering those services. [6]

While process owners had authority without accountability, the reverse was true of everybody else. They were accountable for the delivery of services; but they didn't have the power to determine the processes they used to do their jobs.

If these service-delivery groups failed, there was no way to know whether it was due to their own poor performance, or due to a bad process. Nonetheless, they took the blame.

Process owners implemented processes. In fact, they implemented detailed, rigorous processes. They succeeded at their mission. But the organization became bureaucratic, slow, and inflexible as a result. This structure violated the Golden Rule.

The Golden Rule

The Golden Rule is simply this: **Authorities and accountabilities must match.**

If ever authorities and accountabilities are separated, serious problems inevitably arise:

- On one hand, the person with authorities, but not matching accountabilities, becomes an unconstrained tyrant.

   This wording may seem a bit strong. But in fact, such people can make decisions without bearing the consequences; they can tell others what to do, while others are held accountable for results. So there's little to stop them from issuing edicts that may not be practical, and then letting others take the blame when their commands backfire. Without checks and balances, they do as they please.

- On the other hand, those with accountabilities, but insufficient authorities, are disempowered and can't get their jobs done.

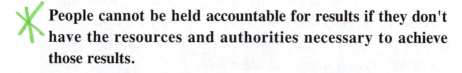
People cannot be held accountable for results if they don't have the resources and authorities necessary to achieve those results.

Sure, you could try to hold staff accountable for things they can't control; but this won't lead to good performance. All it does is establish a scapegoat to blame when things go wrong.

Over time, disempowered people adopt a helpless "victim" mentality, take no initiatives, and spend a lot of time reading Dilbert and laughing about how futile it is to try to accomplish anything important.

The importance of matching authorities and accountabilities has long been recognized. [7] But this Principle doesn't require empirical proof; the logic is compelling: If you aren't equipped to do your job, you can't do your job. And if you meddle in another's decisions, you may be culpable for his/her failures.

Here's a classic case of a violation of the Golden Rule:

In many companies, budgets are determined by the Chief Financial Officer. The CFO is held accountable for the company's financial targets, but not for business results. To achieve his/her goals, the CFO cuts others' budgets, with little regard for the impacts on the business. As you might guess, the CFO is rewarded for cost savings, and not blamed for the poor performance he/she causes throughout the rest of the company.

Here's yet another case from an industrial products company:

The customer-service manager was held accountable for resolving problems, not just coordinating resolution. But he didn't have the resources or authorities to solve many problems. Meanwhile, those who created the problems (production, shipping, billing, etc.) were not held accountable for remediating them. As a result, resolution was impeded and customers' satisfaction suffered.

In a healthy organization, *everybody* is a "process owner" for his/

her own products and services. (Part 8 discusses how cross-boundary processes are optimized in an empowered organization.)

Everybody is accountable for frugality, not just the CFO.

And *everybody* is accountable for resolving any problems they create, not just the customer-service group.

Empowerment

The Golden Rule is the essence of empowerment. **Empowerment means that everybody is accountable for their own results, and has authority over all the information, resources, and decisions they need to do their jobs.**

Empowerment does not mean anarchy. It's not a blank check whereby people can do whatever they please. It simply means that with accountability comes an equal measure of authority (and freedom).

People must be given no more authorities than are warranted by their accountabilities. It's is a fixed-sum equation. If anyone has more authorities than accountabilities, somebody else has less (or they fight for control).

Empowerment isn't just a fad or a nicety; it's an absolute necessity in today's competitive business environment. Organizations can't afford to waste one iota of the talents its people have to offer. Every bright mind has to be engaged in achieving success.

Empowerment is also fundamental to employee motivation. People want to feel proud of their work. And it's hard to feel good if you're treated as a child and told what to do all the time.

At Every Level

Empowerment is not limited to senior managers. Everyone — in every job, at every level, in every corner of the organization — should be empowered to manage their piece of the business, however large or small that piece may be.

But what if your subordinates are inexperienced, and aren't ready to take on very much on their own?

The answer is not to disempower them by telling them how to do their jobs. Rather, it's to empower them in smaller chunks — a series of small assignments, each within their capabilities. This way, even relatively junior people can be empowered with authorities that match their accountabilities, and both equipped and motivated to succeed.

Mandating "How" Instead of "What"

One important implication of the Golden Rule is that organizations should **hold people accountable for results, and leave them free (within bounds) to decide how to attain those results.**

A poignant case example demonstrates the cost of ignoring this Principle and mandating the "how" as well as the "what": [8]

Wallace was an IT consultant. On his last two projects — major ERP implementation programs in two different big companies — he claimed that close to one-third of the costs of these projects was wasted in the name of Sarbanes-Oxley (SOx) compliance.

That's a 50 percent increase in project costs over what they otherwise would have been! I was skeptical. Sure, there are costs of compliance with regulations. But "wasted"?

Wallace insisted he wasn't talking about the costs of complying with SOx. Rather, he was saying that these companies could have complied to the same degree at a much lower cost.

He explained that one SOx requirement is documentation of IT's development and testing processes. These documents are additional work-products. And of course they have a cost.

Documenting processes isn't a waste. The waste resulted from a requirement to do all IT documentation in a specific way. Both of these companies mandated the use of a specific documentation method. (They required UML, Unified Modeling Language, which is an excellent way to describe a system's specifications.)

But the problem was this: ERP systems are well documented by their vendors. The documentation required by SOx would have been standard fare, especially since much of it is supplied by the vendors. But vendors' documentation doesn't fit neatly into the required format. Developing documentation in this specific format was extremely time consuming.

IT leadership drove costs sky-high when they issued an across-the-board edict that dictated the "how" as well as the "what."

Another common example of disempowerment is requiring staff to use a particular project-management method. A standard method may be too cumbersome for small, agile projects, or insufficiently robust for really complex projects. So if people perform poorly, is it their fault, or the fault of the imposed method?

Note that it's not disempowering to clearly define every aspect of the "what," including artifacts like documentation and project status reports. But it is disempowering to require a specific "how."

HOW ~~O~~ WHAT!

So What's Left for Leaders To Do?

As a leader, you may be wondering what's left for you to do if all your staff are empowered. The answer is, a lot!

The Golden Rule doesn't preclude managers from making decisions; they have that authority because they have ultimate accountability for the performance of their groups.

Here are some specific authorities (and accountabilities) that remain with leaders in empowered organizations:

- **Decide the rules of the game**, i.e., create the organizational ecosystem, including the group's structure, resource-governance processes, and culture. And once the right structure is in place, leaders adjust domains as needed, arbitrate boundary disputes, and ensure that teamwork is occurring as it should.

- **Develop talent** within the group. This includes recruiting, inspiring, and coaching subordinates. But coaching is offered as advice which staff may or may not choose to follow. If staff are forced to follow their manager's "suggestions," then that manager must share accountability for their results.

- **Manage performance**. This includes negotiating staff's objectives, giving frequent feedback, measuring results, deciding rewards, and managing performance problems.

- **Manage commitments and resources**. This includes assigning work within the group.

- **Coordinate shared decisions** within the group. Examples include decisions on common methods and tools, and agreements on professional practices.

- **Make decisions** where consensus cannot be reached.

- **Guide business and product (technology) strategies** within the group (recognizing that within the group's overall strategies, entrepreneurs at the next level are empowered to determine their own strategies).

- **Serve as the diplomat**, representing the group to its peers and superiors.

A leader's job is to create the right environment, and do everything possible to help subordinates succeed.

And if you're wondering whether you'll still be in control once you've empowered your staff, the answer is, *absolutely*! In fact, managing by results can give you more, not less, control. Your focus moves up from micro-managing subordinates to the more leveraged level of orchestrating the right results.

Referring back to the metaphor of organizations as machines (Part 1), instead of being a cog in the machine, you'll be the driver of the machine. Working at a more strategic level is good for the organization, and for your career.

Implications for Structure

The Golden Rule affects the design of an organization's culture, structure, resource-management processes, and metrics.

Specifically with regard to structure, the Golden Rule tells us this:

1. **Define jobs based on what people produce** — not by how they do their work (processes, methods, tools, or tasks). Rather than language such as "responsible for doing" (tasks),

job descriptions should be phrased in terms like "accountable for delivering."

2. **Never create a job whose purpose is to disempower others.**

   Don't create steering committees or process owners ("integrating managers" [9]) who constrain others' prerogatives. If you're concerned about teamwork, a far better answer is to install systemic processes that align everybody's priorities (Chapter 49) and link them on teams (Part 8) without disempowerment.

---

*Down-side of violating Principle 1:
unconstrained controls, wasted talents, focus on tasks instead of results, less innovation, inefficiencies and ineffectiveness, unfair scapegoating, unclear accountability for results.*

SYNOPSIS

» The Golden Rule: authorities and accountabilities must match.

» Manage by results, not tasks.

» Define jobs based on what people produce, not by how they do their work (processes, methods, tools, or tasks).

» Never create a job whose purpose is to disempower others.

Chapter 4:
Principle 2: Specialization and Teamwork

> *Nothing is particularly hard if you divide it into small jobs.*
>
> Henry Ford

The next story raises a fundamental question: Why do organizations exist at all? We'll answer that question right after this case study.

Case Study: Customer-centric Structure

The CIO in a large insurance company was under pressure. Business executives were complaining about the IT department's opacity, unresponsiveness, and poor understanding of their business strategies. They were frustrated that they couldn't control IT's priorities, and they didn't understand why many of their requests weren't being fulfilled.

There was even a trend toward decentralization. Many business units had started their own little IT groups, which the CIO disparagingly called "shadow IT." (Of course, the shame was his, not theirs. These groups only existed because business units didn't want to do business with Corporate IT.)

In response, this CIO dedicated a group to each client business unit, and divided his engineering staff among them. Each group was relatively self-sufficient, with all the skills needed to deliver any applications requested by its assigned business unit.

These senior managers also served as the primary liaisons to those business units.

Reporting to the CIO were:

- *Janice, Sales and Marketing*
- *Bill, Finance*
- *Henry, HR and other Corporate functions*
- *Eugene, Engineering and Manufacturing*
- *IT infrastructure and operations*

This client-aligned structure was akin to decentralization, but with formal lines of reporting remaining within Corporate IT.

He felt this structure would appease the business units and stave off further decentralization. He also felt it would help him hold his senior managers accountable for business results and client satisfaction, with no opportunities for finger pointing.

(In another company, this same structure was politically driven. [10] *It had just consolidated its IT staff, and owning a group within IT made the business units more comfortable with shared services.)*

For a while, this structure seemed successful. For example, Sales and Marketing clients felt that IT was more responsive to their needs. It wasn't that Janice's staff worked any harder. But since clients understood the limits to their resources (the people in her group) and could control their priorities, they were more understanding when Janice explained that she couldn't do everything they wanted because her group was busy with other projects for them.

Clients also felt that IT better understood their business. Since Janice and her staff worked with the same clients day after day, a better partnership developed.

At first, clients were happy and the CIO felt less "heat" from his executive peers.

But then, dissatisfaction reemerged. Here's why:

The client-liaison function was a part-time job for Janice. But with a big group to manage, she didn't have much time to spend with clients. At best, she provided a point of contact. But she didn't add much value to clients' strategic thinking.

To make matters worse, Janice faced a conflict of interests. While she was able to maintain good relationships, the advice she offered was biased by the capabilities of the applications engineers that happened to be in her group. Regardless of clients' real needs, she only delivered traditional applications.

Her engineering function was sub-optimized as well. Janice needed a variety of technical specialists in her group to satisfy her clients' needs. So did her peers. Each technology sub-specialty was scattered among the various client-dedicated groups.

This limited their professional exchange. When Janice's staff ran into technical challenges, they might not have known that someone in another group had already figured out the answer. Even if they'd heard of others' work, their peers were busy with other priorities. Costs rose and response times slowed as everybody reinvented solutions to common problems.

Meanwhile, there was little impetus for standards. The people in Janice's group built systems that were optimal for their specific clients, not for the enterprise as a whole. When the company sought cost savings through the consolidation ("rationalization") of applications, they found it difficult. Few of the existing applications could be used more broadly.

This structure also reduced the engineers' ability to specialize. Each group had to produce every type of application needed by its clients. With reduced specialization, costs rose, quality suffered, and response times slowed.

Furthermore, the pace of innovation slowed. They couldn't hire an expert in an emerging technology when there was only enough demand for one person to be shared across the whole enterprise. They had to wait until demand grew to the point where each business unit alone could justify the headcount.

As a result, business opportunities were missed. For example, Janice didn't have specialists in web and mobile applications that would have allowed customers to review and edit their accounts. They missed opportunities to build customer loyalty, just as competitors were hot on the trail of the "digital enterprise."

The impacts extended beyond just IT's performance. All clients needed information about money, customers, employees, and products. Over time, this structure led to multiple general-ledger systems and multiple records for the same customer. Synergies were lost as the company lost a single view of its customers, its resources, and its suppliers.

Due to all these problems, this customer-aligned organization didn't deliver the advantages of its size. It performed no better than a number of small decentralized groups. It wasn't long before business unit executives demanded decentralization of the applications engineering function, rightly pointing out that there was little benefit to centralized reporting.

This CIO had failed to gain the benefits of being one organization. In truth, this was an organization that didn't deserve to stay together because it wasn't adding value.

Why Do Organizations Exist?

This case study leads us to consider the question: *Why do organizations exist?*

Or said another way, why would an organization of 10, 100, or 1000 people perform any better than an equal number of individuals scattered around the enterprise or among small companies?

One obvious answer is that people within an organization work toward a common purpose. But is common purpose enough?

For very simple tasks, it might be. 100 individuals picking up trash along a stretch of roadside may perform as well as a 100-person organization. They'd only have to agree on their territories to avoid redundant work.

Now imagine 100 individuals, working independently, each trying to make cars. Each would have to be an expert in virtually every branch of design, engineering, manufacturing, etc. No one person can possibly know enough about all those professions to succeed.

Independent people agreeing to work toward a common purpose solves the problem of *volume,* but not of *complexity*.

Cybernetic Variety

The reason for this is found in the limitations of the human mind. We can think only so many thoughts in a day, read only so much, and know only so much. We have a finite number of "brain cycles" each day.

There's a scientific way to explain this obvious truth. Cybernetics (systems science) is the study of dynamic systems, like the thermostat that turns on your heater when it gets too cold.

Organizations can be viewed as large, complex cybernetic systems. (Appendix 3 has more on cybernetics applied to organizations.)

In cybernetics, the term "variety" has a special meaning. [11] It's the complexity inherent in a system, multiplied by the pace of activities (e.g., volume of work per day):

$$VARIETY = COMPLEXITY \times PACE$$

Variety is a measure of throughput, or bandwidth — the amount and diversity of information confronting an organization at any point in time. As complexity increases, as volume increases, and as time-frames shorten, variety goes up.

Definition of a Specialist

The world around us is swarming with immense variety, way too much for any one person to grasp. We're all limited in our variety-processing abilities.

So what can you do with your finite brain cycles?

Figure 2: The T-shaped Specialist

GENERALIST   PURE SPECIALIST   PRACTICAL SPECIALIST

You have three choices (illustrated in Figure 2):

1. You could know a little bit about everything — a generalist, the proverbial "jack of all trades, and master of none." You'd be able to do many things, but none particularly well. You'd never know enough about any one field to master it. This describes many individuals working independently.

2. You could focus all your variety-processing abilities on one topic, and learn virtually nothing about anything else. You'd excel at that one thing; but you wouldn't be able to work well with others. This degree of specialization just isn't practical.

3. You could focus on one profession, while still using some of your brain-cycles to know a little about lots of things so as to be able to work well with others. This "T-shaped" person is the definition of a *practical* specialist. [12]

Cybernetics tells us that **people can only be world-class experts at one thing at a time.** We may rotate among specialties in the course of our careers. But at any point in time, each of us must concentrate on a single profession to master it.

Only T-shaped people can be really good at what they do.

An expert is a person who has made all the mistakes that can be made in a very narrow field.

Niels Bohr

Why Organizations Exist

Now back to the question, why do organizations exist? It's simple: to allow people to specialize. Then, when T-shaped specialists in a diverse range of fields work together, an organization has both depth and breadth. It can process more variety.

On the other hand, an organization of generalists performs no better than an equal number of individuals.

The whole point of forming organizations is to permit people to specialize. Within some practical limits, **the more that people specialize, the better they (and their organizations) perform.**

Those who attain to any excellence
commonly spend life in some one single pursuit,
for excellence is not often gained upon easier terms.

Samuel Johnson

Benefits of Specialization

The benefits of specialization have long been recognized. Adam Smith described some of the advantages of the "division of labor" in 1776. [13]

By specializing, people gain deep knowledge of one discipline. They study its ever-evolving methods and tools, and stay abreast of innovations. And by applying their discipline many times in many different circumstances, specialists accumulate experience. They get really good at what they do.

The benefits of specialization are essentially the benefits of professional excellence, including these:

- **Productivity:** Specialists are more efficient. Their increased productivity translates into cost savings.

- **Speed:** With all their experience, specialists have ready answers and don't have to climb the learning curve with each new challenge. And they know the latest methods and technologies in their field, which can leverage their time. Things get done faster.

- **Quality:** Specialists produce higher quality, because they know how. Their products are more usable, more capable, more maintainable, and have lower life-cycle costs.

- **Risk:** People with more competence and experience deliver results more reliably, and with fewer risks.

- **Innovation:** Specialists can keep up with the literature, and they're the first to learn about emerging technologies and techniques. As a result, the pace of innovation improves.

- **Reduced stress:** Specialists experience less stress. They may be under pressure to produce a lot, but they're confident of their abilities. So they're more productive, stable, and happy.

- **Motivation:** With their greater competence, it's easier to excel and rise up the ranks in their fields. And since specialists are more valued in the market than generalists, they have better career opportunities.

If you have any doubts about the importance of specialization, imagine that you need heart surgery.... How do you feel about your general practitioner doing it?

Specialists always outperform generalists.

---

*Down-side of violating Principle 2:
lower productivity, slower delivery, lower quality,
greater risk, less innovation, more stress, lower motivation.*

Concerns About Specialization

Of course, there are practical limits. Organizations can't afford to specialize to the point of becoming "one deep," i.e., dependent for critical services on just one individual. That person may become a bottleneck; and there are problems when he/she takes a vacation or leaves the company.

Sometimes, cross-training can address this problem; in other cases, the structure must define domains more broadly (sacrificing some specialization) to avoid groups of one.

Another concern is job narrowing, where people do the same thing, day after day — like on an assembly line where people specialize in a specific set of tasks. Narrow jobs are disempowering, waste talent, and demotivate staff. [14]

But healthy specialization is not job narrowing. Specialists *do everything* in their field of study. They work with their customers; deliver today's work; support their past work; keep up with innovations; plan and develop future offerings; and improve their processes. There's as much job diversity in doing all that within one specialty as there is in doing a little bit in many fields.

To the skeptics, remember who makes the most money in the real world: The market rewards specialists, not generalists!

Even so, there are some who resist specialization to avoid accountability for excellence in any one thing. They would be best suited to very small organizations, ideally "organizations" of one.

Specialization Requires Teamwork

Others may resist specialization to avoid depending on others, perhaps because they've had unhappy experiences with teamwork. They argue that their mediocrity is good enough, and it's certainly better than depending on others who might cause them to fail.

Indeed, this is the catch....

Imagine an organization made of specialists, all very good at their respective professions, but each acting independently without teamwork. How well would this organization perform?

Terribly! It would perform worse than a collection of independent generalists. Since no one specialist sees the big picture, collectively they'd be unlikely to get anything done.

So the catch is: The more people specialize, the more they become interdependent. Put simply, **you can't specialize if you can't team.**

This is often the gating factor. If an organization isn't good at cross-boundary teamwork — assembling just the right mix of specialists onto teams, and getting them to work well together — then it must back away from specialization so that groups are independent rather than interdependent. That is, it must sacrifice performance to avoid teamwork.

And that leads to a "silo" organization.

The Silo Organization

Some leaders believe that the only way you can get people to work well together is to put them under a common boss, who presumably will form teams and force staff to cooperate. Citing the Golden Rule, they add that to hold managers accountable for results, you've got to give them authority over all the needed resources.

Essentially, they accept the current, limited teaming capabilities of their organizations, and design the structure to minimize interdependencies and "hand offs." They put all the skills needed by a function under its manager. [15]

This is sometimes called the "silo" organization, since it's made up of groups that are self-sufficient and don't need to work together all that much — like a bunch of vertical silos that never touch.

In such an organization, each profession is scattered among the various groups that need it. This fragmentation reduces specialization, because each member of that profession within each silo must cover the entire profession. The result is lower performance; the benefits of specialization are lost.

You don't need to build silo structures to match authorities to accountabilities. Doing so is like saying that you need to own your own grocery store to control what you eat! Just as managers have authority over external vendors, they can manage team-members in other groups who are assigned to them.

The antidote to silos is effective teamwork across boundaries.

> *All things will be produced in superior quantity and quality, and with greater ease, when each man works at a single occupation... without meddling with anything else.*
>
> Plato

Implications for Structure

Here are the implications of our finite variety-processing abilities:

1. The more people specialize, the better they perform. Don't design jobs for generalists, or expect people to be experts at too many things. Jobs should be designed around well-focused specialties.

2. You can't specialize if you can't team well across boundaries. The more you're willing to invest in cross-boundary teamwork, the more you can specialize (avoid silos of generalists) and the better your organization will perform.

The bottom line is, **organizations exist in order to allow people to specialize, which in turn depends on effective teamwork.**

SYNOPSIS

» The more an organization induces specialization, the better it performs.

» You can't specialize if you can't team well across boundaries.

» Don't build silo structures. Instead, invest in teamwork.

Chapter 5:
Principle 3: Precise Domains

The business of everybody is the business of nobody.

Thomas Babington (First Baron Macaulay)

An organization chart determines everybody's "domains" — the boundaries within which each group functions.

Sometimes a group's domain is defined only by a few words in a box. When different people interpret those few words in different ways, three problems occur: overlaps, gaps, and a lack of focus.

Here's one example.

Case Study: Vague Domains — IT Infrastructure versus Enterprise Architecture

A CIO put Rick in charge of the "Infrastructure" group, and Charlie managed "Enterprise Architecture" (EA). There were other managers, but let's just focus on these two:

- *Rick, infrastructure*
- *Charlie, enterprise architecture*

What do those titles really mean?

Charlie (Enterprise Architecture) believed his job was to design the infrastructure. He proposed cutting-edge technologies that weren't yet stable, but constituted an elegant design.

Rick, of course, fought that. He looked bad when he resisted innovation. But he knew he couldn't deliver reliable services with technologies that were not ready for his production environment. So he ignored Charlie and designed his own infrastructure.

It wasn't that Rick and Charlie disliked one another. They fought because they were paid to fight. They both believed they had authority over the design of the infrastructure.

Overlaps

A lack of clarity may cause multiple groups to think they're responsible for the same function — overlapping domains. When domains overlap, staff fight and compete with one another.

Some leaders even create overlapping domains intentionally, thinking that internal competition will elicit better performance. They might also argue that it maximizes creativity, with different alternatives coming from the various competing groups.

In fact, internal competition should never be needed. Most companies face competition. And in monopoly, not-for-profit, and government organizations, external metrics such as customer satisfaction and cost benchmarks can keep people sharp.

For internal service providers, there's competition from vendors. Nothing is sacred. Product development, manufacturing, sales, and all the support functions can be outsourced.

Another very real form of competition for internal service providers is decentralization, where business units have their own support functions rather than work with a shared-services provider.

The truth is, you don't need overlapping domains to motivate performance or creativity. If the need for competitive performance isn't clear to staff, an outsourcing or benchmarking study can serve as a wake-up call, and provide metrics of staff's competitiveness.

Meanwhile, the **costs of overlapping domains** are high:

- **Reduced specialization:** Splitting a profession into multiple groups reduces specialization. Consider two competing engineering groups, each doing everything, versus one specializing in one branch of engineering and the other in another. The costs of reduced specialization, as per Principle 2, include lower productivity, slower delivery, lower quality, greater risk, less innovation, more stress, and lower motivation.

- **Redundant efforts:** More than one group does the same research, and produces the same kind of products. Solutions are reinvented rather than reused. Redundant efforts waste time and money, increasing costs.

- **Less innovation:** When two groups study the same thing, something else that one of them might otherwise have explored is missed.

- **Confusion:** When two groups offer the same service, customers don't know where to go for what, and the organization appears confused and inefficient.

- **Product dis-integration:** Multiple, incompatible versions of the same product can undermine enterprise synergies. For example, in engineering, overlaps may results in dozens of different fasteners (nuts and bolts) which have essentially the same purpose, increasing costs of manufacturing, inventories, and support.

- **Less teamwork:** Internal competition may be friendly, and it may be limited to just part of staff's work. But the inevitable friction undermines teamwork. No amount of team-building can overcome this force built into the structure.

- **Lack of entrepreneurship:** Whenever a single line of business is fragmented, no one feels "ownership" as the entrepreneur who's accountable for its performance, now and in the future.

Whether by deliberately pitting people against one another or inadvertently leaving boundaries vague, internal competition is costly.

A good structure defines **distinct domains**, with no overlapping boundaries.

---

*Down-side of violating Principle 3, Overlaps:
reduced specialization, redundant efforts, less innovation, confusion, product dis-integration, territorial friction, lack of entrepreneurship.*

Case Study: Gaps — Layers of Engineering

Another case study illustrates a second consequence of vague domains: gaps.

In many fields, there are multiple layers of engineering, each building on the lower layers. What happens when layers are missing from the structure?

One IT department had applications development groups, but no groups dedicated to lower-layer technologies (other than the infrastructure managed by Operations).

Chapter 5: Principle 3: Precise Domains 49

Jim was given responsibility for manufacturing and supply-chain systems. While working on just-in-time inventory management, he studied electronic data interchange (EDI). And as he automated the factory, he learned a lot about document management.

Meanwhile, Donna's financial applications group developed expertise in forecasting algorithms and reporting tools. And Carey's customer applications group learned to quickly deliver small web applications and to analyze the masses of data the company collected on customers.

The IT department's structure looked something like this:

- *Jim, manufacturing and supply-chain applications (plus electronic data interchange and document management)*
- *Donna, financial applications (plus forecasting models and reporting tools)*
- *Carey, customer applications (plus Agile development methods and big-data analytics)*
- *Operations*

They were encouraged to share their knowledge with one another. But this rarely happened. Managers allocated all their resources to their own projects. When Jim asked Donna for help with a user-friendly reporting tool for inventory data, her staff were too busy developing financial applications to work on his projects.

Whenever a needed technology was missing (or just out of reach), people muddled through. Multiple learning curves and replication of efforts were costly; and working outside one's expertise eroded performance, quality, and reliable delivery.

Furthermore, no single leader had the job of ensuring that a comprehensive range of supporting technologies was available. So some critical new technologies were ignored, and business opportunities were lost.

For example, Donna knew she should have explored cloud computing and mobile devices; but her financial applications ran on the mainframe, and she had no time to explore other platforms.

This case study illustrates another (opposite) effect of vague domains: Gaps occur.

Gaps

When a function is missing, the results are easily anticipated:

Unreliable delivery: With no one thinking about a function on a daily basis, it's "catch as catch can" — done when the need becomes urgent and obvious, or when people have spare time and happen to think of it. It's not a reliable process.

Reduced specialization: Staff filling a gap outside their specialty aren't experts; and they don't have time to learn that other profession. The costs of reduced specialization were described under Principle 2: lower productivity, slower delivery, lower quality, greater risk, less innovation, more stress, and lower motivation.

Overlaps: Gaps create overlaps. A group that needs a missing specialty fills the gap, but only for itself since this is not its primary mission. In time, multiple groups fill that same gap, creating overlaps, with all the problems described above.

---

*Down-side of violating Principle 3, Gaps:
unreliable delivery, reduced specialization, overlaps.*

Chapter 5: Principle 3: Precise Domains

Focus

People naturally want to feel proud of their work. But when domains are unclear, staff don't know what they're supposed to be good at. Their sense of direction becomes foggy, and their natural drive for excellence is thwarted. They may try to be good at too many things and become generalists. Or they may just follow their personal interests, and gaps or overlaps occur unpredictably.

Clear domains give people the focus they need to excel. They tell people what literature, which conferences, which methods and technologies, and what research to study. Staff develop a feeling of pride in their group's unique function, and strive to do it well.

Clear domains are also the basis for effective performance management. Performance objectives and other metrics can be clearly defined only if jobs are clearly defined.

Beyond the individual level, an organization as a whole requires clear domains to operate as planned. Both customers and staff need to know where in the organization to go for the help they need.

In summary, a good structure provides **crystal clear definitions of each group's domain.** The entire scope of the organization should be divided among the groups within it, **with no gaps or overlaps, and with clearly worded boundaries.**

Good fences make good neighbors.

Robert Frost [16]

Case Study: The Evils of Roles and Responsibilities

Even if boundaries are clear, the wrong kind of language can hamper performance.

Traditional "roles and responsibilities" blur accountabilities and authorities. Here's an example from an IT organization where the applications-development process was very confused.

Ostensibly, Aaron ran the applications engineering group.

But "business analysts" in Mark's group worked with clients to define requirements and high-level designs.

And "project managers" reporting to Jay (the PMO) were responsible for large strategic projects.

    └ *Aaron, applications engineering*
    └ *Mark, business analysts*
    └ *Jay, project management office (PMO)*

There was a lot of friction, so the CIO instructed them to work out their respective "roles and responsibilities." The answer they came up with was something like this:

- *Aaron was responsible for the applications-development process. Jay decided project-management methods. They promised they knew the difference.*

- *Aaron defined projects for his group. But Mark's analysts were responsible for specifications and high-level designs.*

- *Once specs and high-level designs were done, Mark handed projects off to Jay's project managers (if they were big and*

Chapter 5: Principle 3: Precise Domains 53

"strategic") or to Aaron's group (for smaller projects). Jay and Aaron were both responsible for project delivery.

- *Even for the big projects that Jay's PMO managed, Aaron had a role. He supplied engineers to do most of the work.*

- *Aaron retained responsibility for maintenance and support of applications once they were deployed.*

Confused? So was the CIO. He asked a simple question: "Who's responsible for project success?"

Mark looked away; it wasn't his problem, even though his analysts controlled a critical step in the process.

But both Aaron and Jay raised their hands. If it's a small project (such as an enhancement to an existing application), Aaron's group was responsible. Anything big was deemed a "strategic project" and Jay's PMO took over.

Delivery problems persisted. Neither Jay nor Aaron could control the resources they needed to get the job done. Jay pulled engineers out of Aaron's group to staff his strategic initiatives. And Aaron yanked engineers off Jay's projects to deal with urgent maintenance issues, putting Jay's projects in jeopardy.

Even small, urgent projects took a long time. Mark chose the requirements-planning method. Aaron chose a meticulous development method which slowed projects, no doubt for good reasons; and Jay decided the project-management method (which also slowed projects). These three methods didn't mesh, and required lots of paperwork for every hand-off. All three managers shared responsibility for the overall process, but none had the power to fix it.

And Aaron's engineers, responsible for small projects, got little support from Jay's PMO staff who were busy with their big initiatives. So their project-management capabilities remained weak.

Finally, Aaron was held accountable for the long-term integrity of applications; but he couldn't control all the decisions that affected it. They <u>all</u> made decisions that impacted the quality of applications.

They were all set up to fail, with overlapping accountabilities and without the authorities they needed to deliver results. Even though they'd divided up the tasks, they stepped on each other's toes and tensions rose.

RACI

A more sophisticated framework for defining roles and responsibilities is called "RACI." [17] For all the tasks in a process, leaders sort out who has:

- **R**esponsibility for doing it,
- to whom they are **A**ccountable,
- with whom they have to **C**onsult, and
- whom they're to keep **I**nformed.

The RACI method has all the same problems:

- **Responsibilities:** Like simple roles and responsibilities, it defines accountabilities for tasks and processes, not results.

- **Accountable:** This person with authority to make or approve decisions may or may not be the customer; this risks separating authorities from accountabilities.

- **Consulted:** RACI defines key contributors, but not their deliverables; so they lack accountability for results. And this fixed list of contributors may not be right for every project.

- **Informed:** Again, a fixed list may or may not be the right stakeholders for every project. This is bureaucratic, not flexible and dynamic. People spend a lot of time "consulting" and "informing," whether or not it adds value.

Downside of Jobs Designed Around Tasks

If jobs are designed around tasks (or roles and responsibilities) rather than results, many dysfunctions occur:

- People execute tasks. They're not asked to think creatively about how to attain intended results. This wastes the insights of those in the best position to know the best way to get things done. And tedious jobs lead to boredom and degrade morale.

- Each group sub-optimizes (and perpetuates) its steps in the current process. And if the tasks don't add up to the intended results, they're not in a position to fix it.

- The boss who coordinates all the tasks must plan every project in detail, monitor everyone's work, and adjust tasks when things go wrong. Busy managing tasks, the boss doesn't have time to think about his/her own challenges such as building relationships, innovation, and business strategies.

- If a project requires a new task, it may be nobody's job and hence the task may not get done. Of course, this puts the ultimate deliverable in jeopardy.

- When disempowered people interact with customers, both parties will be frustrated. Customers will ask for results that task-focused staff are not able to offer. And staff will feel badly that they can't satisfy their customers.

- People believe they've done enough if they've put in the required hours and completed their tasks, whether or not the job is done. They go through the motions, without caring (or even knowing) whether those motions produce the intended results.

- When sorting tasks among groups, there's nothing to stop you from accidentally separating accountabilities and authorities.

- More broadly, roles and responsibilities don't define accountabilities for leadership — for improving methods and tools, sorting out processes and relationships with peers, exploring emerging technologies, selecting and managing vendors, and developing strategies. It doesn't appoint leaders responsible for managing today's business and for planning tomorrow's.

This all adds up to unhappy customers and staff.

The problem with "roles and responsibilities" is, you can sort tasks in excruciating detail, and you still won't know who's accountable for delivering the organization's products and services.

Domains Based on Results

Organizations don't make money by going through the motions. They make money by producing results. Therefore, the key to good domains is **bounding what people produce** rather than what they do.

Dividing up results is actually a lot easier than sorting tasks. The list of an organization's products and services is far shorter than all

Chapter 5: Principle 3: Precise Domains 57

the tasks people do and roles they play. And by defining who *produces* what, you'll automatically know who *does* what.

Let's try this results-oriented approach in the IT organization described above:

Aaron is the applications manager. He's accountable for the entire portfolio of applications. That's that. But now let's consider what help he might need....

If Aaron isn't clear about exactly what a customer needs, he may get help from Mark's business analysts. Mark's group produces requirements definitions, not product designs (not even at a high level, since design is an engineering task).

Once requirements are clear, Aaron runs the project. He produces anything related to applications engineering — a repair, an enhancement, even major strategic initiatives. He's the single point of accountability for all applications.

Jay is an expert in project management, but that doesn't give him the right to take over Aaron's projects. Jay sells Aaron project plans, training, project facilitation, and project administration services. But whether or not Aaron's engineers get help from Jay, they're still accountable for delivering projects, big and small.

It's Aaron's job to stay abreast of innovations in engineering methods. Meanwhile, Mark continually refines his requirements-planning methods. Jay runs a consulting business and has to keep up with project-management methods and tools.

Since each group only produces results within its domain, boundaries are clear. Roles and responsibilities (and hand-offs) are clear because everybody knows what they produce, for clients and for one another.

Implications for Structure

Precise domains, defined by results, help people understand how the structure is supposed to work. Clients know whom to call for what. Staff better understand what's expected of them.

Also, teamwork improves because everybody is empowered to do what it takes to deliver their products and services, and they know where to go for help when they need it.

The importance of precise domains can be summarized in four implications for structure:

- Provide every group with **precisely worded boundaries.**

- Ensure **no overlaps.** Never create internal competition.

- Ensure **no gaps**. Place every needed specialty within the domain of one group.

- Domains should **bound what people produce,** not the roles they play or the tasks they're responsible for.

Clear domains take more than a few words in a box on an organization chart — vague phrases like "infrastructure" or "operations." Each box should be accompanied by a carefully worded paragraph that clearly bounds what each group produces.

Chapter 5: Principle 3: Precise Domains

SYNOPSIS

» Provide every group with precisely worded boundaries.

» Ensure no overlaps to avoid internal competition.

» Ensure no gaps, with every needed specialty within the domain of one group.

» Domains should bound what people produce, not the roles they play or the tasks they're responsible for.

Chapter 6:
Principle 4: Basis for Substructure

Let's say an organization is big enough to require multiple groups in essentially the same profession. In large organizations, for example, there may be multiple engineering groups.

Principle 3 says that their domains are defined by their respective results. But how do you divide all the various deliverables among those multiple groups?

The next case study shows how critical this decision is.

Case Study: Structure by Clients' Business Processes

A CIO viewed IT as a means of supporting clients' business processes. This was a rather limited view; but at least he wanted to focus staff on business rather than just technologies.

Based on his view, he dedicated a group to each of the company's core processes. Gail was given responsibility for the corporate product-development process. Jerry looked after the entire order-to-invoice process. And Tom supported all the company's financial and administrative processes.

- *Gail, product-development process*
- *Jerry, order-to-invoice process*
- *Tom, financial and administrative processes*

Each group was expected to become familiar with the workflows, and design solutions that optimized these business processes.

In practice, this structure produced some serious problems:

Redundant learning curves: *The various processes all required many common skills. This caused a lot of parallel learning curves.*

For example, under the influence of innovative clients in R&D, Gail was the first to develop applications on then-new mobile devices. Later, when Jerry embarked on a sales-force automation project, his group had to research the same technologies.

While they helped one another with advice, they weren't able to flexibly deploy existing expertise to one another's project teams. So staff still needed time for learning what others already knew, which increased costs and delayed projects.

Redundant work: *Parallel efforts led to redundant work. For example, both Gail and Jerry developed document-tracking modules. This drove costs up further.*

Reduced specialization: *Skills needed by multiple processes were scattered among the groups, reducing staff's ability to specialize.*

For example, Gail and Jerry both needed expertise in product data. Jerry and Tom both worked on customer, sales, and financial data. And all three worked on human resources data. They became specialists in clients' business processes, but generalists with regard to their own profession of engineering specific types of applications. Performance naturally suffered.

Product dis-integration: *Each group independently developed its own versions of financial, customer, and product databases. Fragmented data caused confusion when different reports gave clients conflicting views of the "same" data.*

When they did collaborate on a few common applications and databases, other problems occurred. They each enhanced the same applications at various times. As a result of "patch on patch," systems integrity deteriorated.

Weak client relationships: *The CIO expected that this structure would bring IT closer to clients. However, this was not evident.*

Most business units were involved in numerous business processes. For example, Manufacturing was involved in both product development and production. Thus, both Gail and Jerry served Manufacturing executives, focusing on different aspects of their jobs.

With no single point of contact for a client like Manufacturing, it appeared that the various IT groups were competing for the clients' attention. IT appeared poorly coordinated.

This also meant that clients had to determine which workflow they wished to discuss before they could know whom to call. As a result, IT's client liaisons were not included in clients' thinking early on, where the real strategic discoveries are made.

Furthermore, since processes touched multiple clients, IT managers couldn't focus on just one business unit. With more business units to cover, they had less time to get to know the people in each. This also distanced IT from clients.

Finally, some clients weren't involved in any of the core business processes. This doesn't mean that they weren't important. For example, top executives are rarely engaged in routine business processes. Clients such as the president and the executive vice president of planning and business development had no designated client liaisons, and received poor service.

Biased business diagnosis: *A group dedicated to a business process is naturally biased in favor of automating that workflow.*

Gail, for example, was paid to believe that automating the product-development process was the most important thing to do.

However, the corporate engineering function might have benefited far more from solutions that improve the effectiveness of key individuals — for example, engineering design tools — solutions that have nothing to do with automating the workflow.

Similarly, a Manufacturing executive might be involved in a critical decision, like consolidating plants, that requires decision support or collaboration tools — challenges that have nothing to do with any of the chosen processes.

Due to the bias for automating business processes built into this structure, high-payoff opportunities were missed.

Basis for Substructure

The way you divide domains among groups within a profession is termed the *basis for substructure*. [18] It determines people's specialties — their bottom-of-the-T. For example:

- If you assign groups to clients' business processes, they'll become experts on those workflows, but generalists with regard to clients, their engineering disciplines, and services.

- If you assign groups to clients (for internal service providers, business units; for companies, territories), they'll become very close to those clients while becoming generalists with regard to the organization's products and services.

- If you assign groups to technologies and disciplines (e.g., specific products), they'll become experts in those products while becoming generalists with regard to the clients.

- If you assign groups to services, they'll become experts in the delivery of those services, while becoming generalists with regard to the organization's clients and its products.

What Happens If You Get It Wrong

As the above case study illustrates, the consequences of choosing the wrong basis for substructure are serious:

Reduced specialization: If the basis for substructure is anything other than a function's expertise, then staff will specialize in something other than their primary mission.

In the case study, groups specialized in business processes, not in knowing clients or in their engineering profession.

As another example, when Sales is substructured by product lines (a sales force within each business unit), specialization in the clients' business is reduced. Since each Sales group must cover all the clients, staff have less time with each client, and get to know them less well.

As per Principle 2, when specialization is reduced, performance suffers: lower productivity, slower delivery, lower quality, greater risk, less innovation, more stress, and lower motivation.

Domain overlaps: At the same time, inappropriate substructure often creates overlapping domain.

For example, if Engineers are divided into client-dedicated groups, multiple groups deliver essentially the same solutions to their respective clients. So there's often a lot of reinvention.

And when Sales is substructured by product line, clients are

Chapter 6: Principle 4: Basis for Substructure

confused by multiple points of contact, where they have to call one person for some products and another for others.

The costs of overlaps were described in Principle 3: reduced specialization, redundant efforts, less innovation, confusion, product dis-integration, territorial friction, and lack of entrepreneurship.

Inappropriate biases: The wrong basis for substructure can also induce the wrong biases. As a result, staff give poor advice, and optimize the wrong objectives.

For example, a Sales force that's divided by product lines can't be a trusted advisors to clients. Their recommendations are always biased toward their own product lines, not clients' real needs.

Disempowerment: Another problem that occasionally results from inappropriate substructure is a violation of Principle 1, the Golden Rule: Two groups may fulfill essentially the same function, one doing the thinking (e.g., planning, or designing processes) while the other does the delivery. Neither is wholly responsible for results.

The costs of this disempowerment are described under Principle 1: unconstrained controls, wasted talents, focus on tasks instead of results, less innovation, inefficiencies and ineffectiveness, unfair scapegoating, and unclear accountability for results.

Down-side of violating Principle 4:
reduced specialization, domain overlaps,
inappropriate biases, disempowerment.

Implications for Structure

The organization chart defines everybody's specialties — their "bottom-of-the-T." People concentrate on whatever their job's domain may be, and they become generalists at everything else.

So the simple guideline is: **Use a basis for substructure that exactly matches what people are supposed to be good at**.

There's no one right answer for an entire organization. But for each specific function, the specialty ("bottom of the T"), and hence the right basis for substructure, should be evident.

For example, if you want engineers to be good at designing things, define their domains by what they design. If the job of Sales is to know clients, define their domains by the clients they serve.

SYNOPSIS

» The way you divide domains among groups (the basis for substructure) tells people what they're supposed to be good at.

» Use a basis for substructure that exactly matches a function's specialty.

Chapter 7:
Principle 5: Avoid Conflicts of Interests

To support empowerment, Principle 3 states that groups' domains should be defined by the results they produce. Of course, an organization produces many different results. So what goes best with what? What's the right way to cluster a set of results into group's domain?

There are two considerations: conflicts of interests (this Principle 5), and professional synergies (Principle 6 in the next Chapter).

To illustrate the first of these two issues, consider how this case study builds conflicts of interests into people's jobs.

Case Study: The Governance and Client Liaison Group

An IT organization found itself overwhelmed with unchecked demand coming from both its clients and internal projects. It needed to better manage the process by which new work was taken in. The CIO also wanted to better align IT with clients' business strategies by setting the right priorities.

Instead of addressing demand management and alignment through business-driven resource-governance processes, this CIO put Martha in charge of a "Governance" group. Her responsibilities included business relationship managers, demand management (deciding priorities among proposed projects), and project portfolio management (a project management office, PMO).

Other groups were responsible for project delivery and ongoing services.

Martha faced many conflicts of interests:

Martha had the authority to decide project priorities. Since not all benefits are measured to the point of calculating accurate returns (ROI), she couldn't just go by the numbers. So she had to judge the merits of clients' projects. This conflicts with her business relationship management role which attempts to be on clients' side of the table. Will clients be open and trust the people who will later judge the merits of their requests?

Similarly, judging internal projects proposed by peers undermines the service orientation that's expected in her project management function, which is supposed to help peers deliver those projects.

And since Martha's job was to limit demand to available resources, will her business relationship managers be aggressive about seeking new high-payoff opportunities (which makes limiting demand more difficult)?

Also, the business relationship managers help some clients discover new opportunities. Will projects that other clients define on their own (without this group's help) get fair treatment when it comes to setting project priorities?

And will that discovery process (which should be business driven and unbiased) recommend services as well as projects, despite the bias toward projects coming from Martha's project-management function?

By the way, some high-payoff opportunities may be urgent. The project-management function isn't going to like these disruptions to its well-planned project schedule.

Types of Conflicts of Interests

If you put the wrong deliverables together in a single job, you can inadvertently create conflicts of interests — two or more missions that are in some way opposed.

There are five fundamental conflicts of interests inherent in all businesses, and in many organizations within enterprises:

- Invention (major innovations) versus operational stability

- Purpose-specific solutions (applications) versus common components that contribute to various purposes

- Enterprisewide thinking versus the focus of a specialist

- Specialization in a subset of the product line versus unbiased diagnosis of clients' needs

- Service orientation versus audit

Invention Versus Operations

Some managers find themselves responsible for both researching and developing new technologies (invention), and also for operating them. Combining invention and operations may be an attempt to give a manager a complete piece of the business to run. It is, however, misguided.

Consider the following case, again from an IT department:

"George, you've really done a great job on that project. I think you're going places around here."

"Thank you, sir!"

"In fact, I'm doing a little reorganizing, and I have a new assignment in mind for you. You know we've been having a bit of trouble getting our infrastructure operations under control. I'd like you to clean that up for me."

"Great, boss! Sounds like a fun challenge. But tell me, who've you got in mind to take over my job of infrastructure engineering?"

"George, you did well a couple of years ago in operations, and you've done well in engineering. At this point in your career, I think you ought to be able to handle both."

"Uh-huh...."

"With the proper headcount, of course."

"Okay, boss!"

6 months later....

"George, can I talk to you for a minute?"

"Sure, boss."

"I'm really pleased to see the way you've stabilized operations...."

"Well, thanks, boss!"

"But I've been getting a number of calls from the other VPs complaining that you've been holding up their projects. You know how I hate calls like that. I'd like to politely ask, what the #@!! is going on?"

"Well, sir, I've been working 60-hour weeks to document procedures. I've still got a ways to go; and I haven't started thinking about capacity planning, change control, disaster recovery, and security administration. Meanwhile, half the day

Chapter 7: Principle 5: Avoid Conflicts of Interests   71

seems to go into fire-fighting. The last thing I need right now is another major application coming on-stream!

"And, boss, every time we bring a new application into production, it disrupts operations. I think we have to slow down and ensure proper change control. I decided that a once-a-month window for installing new applications would be more than ample.

"And another thing.... I can't let the engineers put just anything into production. Then I'd be blamed if it doesn't run reliably. So I've instituted a lot more rigor around testing and quality control — even for small projects like enhancements. I think we've made great progress stabilizing operations. Good stuff, eh boss!?"

"George, George, George.... You know we need to do more than keep existing systems running. We've got to get new applications up. We need to explore cloud computing and social media. We need innovation! What's the problem???"

The problem is simple: Invention and operations are like oil and water — they just don't mix. Here's another <u>case</u>: [19]

A promising executive within a telecommunications cable manufacturer was given responsibility for both business development (acquisitions) and operating acquired companies during their transition into the corporation. After the first major acquisition, the CEO was disappointed to find that little was done in the way of new deals.

Why? Running and integrating the acquired company required this executive's full attention. After the first big acquisition, he had no time for new deals. By combining invention and operations, the CEO was left without a business development function.

The term "invention" is meant to imply creating entirely new things, for example developing new ideas, products, and methods. This is a special type of innovation, beyond the marginal changes of continual improvement.

Invention is the crux of this conflict of interests. Any function can be innovative. But *invention* and operations are antithetical.

When one person is asked to do both, he/she is not likely to find the ideal balance. Operational issues tend to take precedence. Fire-fighting swamps invention. Short-term problems take attention away from long-term opportunities. Those who "keep things running" have little time for future-oriented invention.

It's not just a matter of having the time. Invention is not in their best interests. Major changes threaten the stability and efficiency of operations. Operations staff will pursue innovation on the margin; that is, they'll get better and better at what they currently do. But they're unlikely to pursue breakthroughs which inevitably disrupt operational stability.

Generally, this conflict of interests constrains invention. But there are cases where the opposite is true. Under a leader who prefers engineering, invention takes precedence. Every new idea finds its way into production, whether or not it can be supported reliably. Under that kind of leader, operations never stabilize.

Purpose-specific Solutions Versus Components

Many organizations produce two different kinds of products: some are purpose-specific, and others are components which may stand alone or be part of purpose-specific solutions — complete products, versus the parts that go into them.

Take the construction industry as an example. Some experts

Chapter 7: Principle 5: Avoid Conflicts of Interests   73

design buildings (architects), and other experts design bridges (civil engineers). Both of these purpose-specific engineers draw on experts in common components, such as structural engineers, electrical engineers, and traffic engineers.

In IT, business applications are purpose-specific. One team of developers may specialize in financial applications, and another in customer applications. All the different applications engineers draw on experts in component technologies, such as computing platforms, databases, and middleware.

If a group is expected to produce both purpose-specific solutions and component solutions, an unhealthy bias is built into the organization chart. Staff tend to put everything on familiar platforms.

Conversely, keeping them separate has advantages. Sergio Paiz offers a case example:

"PDC bought a bleach operation with two factories. One factory produces hypochlorite, the base chemical. It's sent to the second factory which dilutes it and blends it with fragrances to produce consumer bleach, an application managed by our Consumer Division. Our Industrial Division was quick to figure out that hypochlorite is used in several industrial applications like mining and agroindustry. Distinguishing purpose-independent (base chemical) from purpose-specific (consumer bleach) products led to a new source of revenues."

Enterprisewide Thinking Versus the Focus of a Specialist

Some decisions — such as policies, plans, and standards — are enterprisewide, affecting many different stakeholders.

If these enterprisewide decisions are assigned to any one of the impacted domains, decisions will tend to be biased in favor of that one stakeholder. The knowledge and interests of other affected groups may not be fairly represented.

Continuing the IT examples, consider decisions on technology standards. Standards affect all the IT applications and infrastructure domains, as well as operational services, customer support, and in some cases even clients.

If network standards are decided by network engineers, they may neglect the interests of other engineers who must design systems to be compatible with the network, and the concerns of the infrastructure services group that's accountable for running the network.

Product Specialization Versus Unbiased Diagnosis of Clients' Needs

Technical excellence requires a deep understanding of a single domain — a subset of the organization's products and services.

Specialization brings quality, but it also brings bias. This bias is not unhealthy, but rather it's a natural outcome of dedicating one's career to a particular specialty.

On the other hand, when diagnosing clients' needs, the organization should be completely unbiased. It must listen carefully to clients' business challenges, and prescribe the most relevant subset of its entire product line.

If staff are expected to do both, a conflict of interests arises. As product specialists, they *should* be biased. But when talking to clients, they're expected to provide unbiased, business-driven advice and "sell" whatever is most needed.

It's impossible for people to be both biased and unbiased. As the old saying goes, "Give a child a hammer, and everything looks like a nail!" Despite their best efforts to be objective, staff see only needs for their favored solutions, and higher-payoff opportunities may be missed.

Service Orientation Versus Audit

The primary mission of most functions is serving others — peers inside the organization, clients throughout the enterprise, or external customers. Customer focus is a key success factor.

But there may also be a need for an "audit" function which judges others, and may even have veto-power over others' decisions. For example, auditors may be necessary to ensure compliance with financial laws, regulations, policies, and ethics principles.

Mixing audit functions with service functions creates a conflict of interests. It's impossible to build a relationship with customers while also judging them. You just can't say, "I'm from the Internal Revenue Service; I'm here to help!"

An IT example is giving the PC group the power to decide whether users need a new PC. Clients wouldn't perceive them as service oriented, and wouldn't openly discuss their real needs.

Consequences of Conflicts of Interests

When structure tells an individual to go in two (or more) opposing directions, there are adverse consequences both for the organization and for the individuals involved:

Gaps: Putting people in conflict-of-interests situations doesn't produce excellence at both missions. With finite brain cycles, people might be mediocre at both missions. More commonly, they prefer one (based on their predilections), and performance at the other mission falls short.

In the example of George above, operations took precedence and infrastructure engineering (invention) became a gap.

Unpredictable balance: The balance between conflicting missions is not a deliberate process, an interplay among peers that analyzes trade-offs. People independently decide the balance based on their limited view of the organization's needs, their intuitions, and personal preferences.

As one manager leans one way and another in the opposite direction, when their decisions are added up, statistics tells us there will be a tendency toward the mean — regardless of the needs of the business at that time.

Meanwhile, the organization's top executive has little control over the balance. If the organization needs to emphasize operational efficiencies to save money, or innovation to reap opportunities, the top executive has no knob to turn. There's no explicit way to adjust the balance on these conflicting objectives.

Stress: A more personal consequence is stress. When people are expected to go in opposing directions, they don't know what to do; and typically they're fully aware of their failure to succeed at both. Highly stressful jobs lead to poor motivation, performance deficiencies, and even health problems (like ulcers).

---

*Down-side of violating Principle 5:
domain gaps, unpredictable and uncontrollable
balance of conflicting objectives, stress.*

Implications for Structure

Since organizations may face all these conflicts of interests, top executives inevitably must cope with these paradoxes and decide the balance point on each. But at lower levels, it's not healthy to combine functions with opposing objectives.

This leads to a simple structural guideline: **Reserve conflicts of interests for the highest possible level in the organization chart,** ideally for the top executive alone.

A healthy structure defines jobs that focus on only one side of each paradoxical dimension. This gives staff clear, non-conflicting job objectives. And it allows the executive to explicitly adjust the balance by shifting resources between groups.

SYNOPSIS

» Don't combine functions that have inherent conflicts of interests.

» Reserve conflicts of interests for the highest possible level in the organization chart.

Chapter 8:
Principle 6: Cluster by Professional Synergies

As you combine your organization's many deliverables into domains, the first consideration is avoiding conflicts of interests. The second is professional synergies.

Here's a case study that illustrates the wrong way to go.

Case Study: Process-centric Groups

With some critical internal processes in need of attention, a CIO tried a process-centric structure. He dedicated groups to each major process within IT to minimize "hand-offs."

The structure included these groups (and a few more):

- *Ruth, applications development process*
- *Bob, infrastructure engineering process*
- *Matt, incident management process*
- *Art, operational service delivery process*

Optimizing processes is a good thing. But this structure created "silo" groups for each process, each containing the various specialists it needed. It scattered the "campus" of similar professionals.

With each group attempting to cover many engineering disciplines, they all became less specialized, and hence less effective. Costs rose, quality suffered, and the pace of innovation slowed.

Furthermore, the structure was disempowering. For example, Art was forced to operate whatever infrastructure Bob developed. He was accountable for service delivery, even though he had only indirect control of his own assets.

Smaller, less-visible processes were not represented in the structure, such as business opportunity analyses and standards planning. Since they were counting on the organization chart to make processes work, these other critical processes languished.

And this structure sent the wrong signals. Staff focused on executing existing processes, not on running businesses and pursuing innovations in their professions.

Finally, illustrating this next Principle, technology experts scattered among the various groups didn't coordinate professional practices such as methods and tools; they didn't share components; and they did poorly at product integration. Both professional and enterprise synergies were lost.

Structuring by internal processes is an extremely costly way to optimize workflows. If you're concerned about the effectiveness of your internal processes, a process-facilitation function can help. And Part 8 explains how to build high-performance, cross-boundary teamwork.

As long as you're willing to invest in processes and teamwork, there's no need to put diverse professions under a common boss just to get them to work together.

Clustering Similar Professions

If you don't have to put people under a common boss to get them to work together, then you're free to **cluster staff by their professions**. This produces many kinds of synergies:

Professional synergies: Working together in the same group encourages professional exchange — sharing experiences, discoveries, refinements of techniques, and best practices.

This reduces redundant learning curves. Everybody is better informed, which improves speed, quality, and innovation. And institutional knowledge is better preserved by concentrating (rather than scattering) it.

Furthermore, similar professionals can share their work products. Even if each project is unique and solutions are customized for each client, there may be opportunities to reuse lower-level modules and designs. At a minimum, staff can make use of others' experiences.

Sharing in any of these forms saves time and money, and common components may improve the quality and maintainability of the organization's products.

Conversely, scattering a profession destroys that "campus effect." People don't get together often enough to learn from one another. They relearn and reinvent. This wastes time and slows delivery. It dampens the pace of learning and innovation. And it reduces depth of expertise. The result is lower productivity and quality, at a higher cost.

Management synergies: A manager focused on a set of similar professions better understands how to manage those specialists. He/she is a better leader and mentor, and can create a sub-culture appropriate to the profession.

Conversely, when a small group in one profession is placed under a manager of a completely different profession, its boss may not understand the function well enough to lead it. Mentoring will be weak, and the sub-culture may be inappropriate.

And with a given profession reporting to multiple managers, there's less management control. For example, it's harder to develop common methods and enforce professional standards.

Workload synergies: A larger pool of staff can better manage workloads. When one person or group becomes too busy, a common manager can temporarily assign people from other closely related disciplines; and since professional skills are similar, there's a reasonable chance that the loaned staff will be productive.

However, when scattered about, a little group of professionals under one manager finds it difficult to manage peak loads. It's hard to borrow people from a group reporting to a completely different manager.

Negotiating power: When a profession is consolidated, its manager has more buying power and can negotiate better deals from suppliers. The group may also be able to save money by sharing tools (e.g., software licenses) and vendor services.

When a profession is divided into multiple groups, each procuring its own tools and services, buying power is diminished.

Career paths: A single, larger group of all those in a given profession offers better career opportunities. Supervisory positions within that larger group may provide promotional opportunities for those who are excellent in the profession.

By contrast, when a profession is scattered into small groups under other functions, staff find it difficult to advance. Positions at the next level up may not benefit from their professional expertise. So when they look for a career path, there's no place to go other than to leave their field of study behind.

Domain adjudication: A common manager overseeing a collection of similar domains resolves boundary issues, and ensures that accountabilities for emerging technologies and disciplines are assigned to one group. This reduces the chances of domain overlaps and gaps.

On the other hand, if a profession is scattered around the organization chart, the groups may only meet at the level of the top executive. This busy leader doesn't have time to personally adjudicate domains, or to ensure that every new sub-specialty is covered somewhere, without overlapping domains.

The result is typically domain overlaps and gaps. Principle 3 described the problems: reduced specialization, redundant efforts, less innovation, confusion, product dis-integration, territorial friction, lack of entrepreneurship, and unreliable delivery.

Simplicity: Putting similar professions in one place makes it easier for others to understand the structure and find the source of needed products and services.

When a profession is scattered, it's harder to know where to go. It creates confusion for clients as well as others in the organization.

Product synergies: When similar professionals collaborate more, the organization's products are likely to be better integrated. This can produce enterprise synergies, e.g., when clients collaborate better by using common tools and services.

For example, when all the IT engineers who work on customer applications are in one group, they're more likely to create systems that allow all the enterprise's business units and functions to share a holistic view of customers.

Alternatively, if the pieces of a profession report to many different managers, there's a tendency for each group to go its own way. For lack of collaboration, an organization's products don't fit well together (product dis-integration). Silo structures tend to produce monolithic products, not a modular, interoperable product line. [20] This increases costs and sacrifices enterprise synergies.

Down-side of violating Principle 6:
less professional exchange, weaker mentoring and management controls, less load balancing, higher costs, limited career paths, domain overlaps and gaps, confusion, and missed enterprise synergies.

Implications for Structure

The simple guideline for structure is: **Cluster domains based on professional synergies** (not who works with whom). Put all the staff who share a common profession together under a common boss.

SYNOPSIS

» If you trust that you'll build processes of cross-boundary teamwork, there's no need to cluster functions based on who works with whom.

» Cluster similar specialties to maximize professional synergies.

Chapter 9:
Principle 7: Business Within a Business

The seventh (and last) Principle speaks to the way you think about domains, and staff's authorities and accountabilities.

A common mistake is to hold people accountable for implementing their specialties. This may sound logical; but it's dangerous, as illustrated in the next case study. For this case study, let's use an example far from the world of IT.

Case Study: Safety Group that's Accountable for Safety

A local water district serving nearly 10 million households employed staff in three shifts to operate and repair its facilities. [21]

Separately, a Safety and Environmental Compliance group did inspections, supervised the handling of hazards, and oversaw the Operations staff while they did the work. For example, they monitored oxygen in man-holes while workers were inside, inspected welders' equipment, and oversaw excavations.

Under the CEO were these two groups (among others):

- *Randy, operations and repairs*
- *Barry, safety and environmental compliance*

This structure didn't work at all well. Here's what happened:

Safety was reduced: *Work crews weren't trained in safety, since the Safety group was supposed to take care of that. But Safety staff couldn't oversee every detail. Worse, sometimes work crews grew impatient waiting for Safety staff to show up, and went ahead* *without them. So mistakes were made, and the number of safety incidents went up.*

After paying a large fine for an environmental accident, the Board hired a respected consulting firm to benchmark the water district against high-performing peers. The study found that they suffered incidents five to seven times the average, and incurred fines that were three to five times the average.

Costs increased: *At that time, the 80-person Safety group was requesting an additional 20 staff. However, that same study found that best practices were just 5 to 10 safety experts dedicated to establishing policies, training workers, and collecting data.*

The economics were clear. A Safety group held accountable for other people's behaviors was less effective and more expensive.

Safety is an attribute of people's work, not a product in itself. *Everybody* must run safe businesses and produce safe products. So the Safety group should have been in the business of providing training and consulting to Operations, not "implementing safety."

Like in any function, *success depends on understanding what business you're in.* And that leads us to the final Principle.

Why Entrepreneurs Love Their Jobs

This final Principle embodies the first six, and takes them to the next logical step. It's simply this: **Each group should be defined as a business within a business**.

Since 1980, I've been an advocate of managing groups within organizations as businesses within a business. [22] My beliefs were validated when I had occasion to meet with some highly successful entrepreneurs. Each led a profitable business, employing from dozens to thousands of people.

Clearly these were very smart and well-educated people who could have been top executives in global companies. But instead, they chose to build their own small businesses. I asked them why. Perhaps surprisingly, it wasn't for the money. Here's what they said they liked about entrepreneurship, sorted into two categories:

Empowerment:

"...*owning my time, choosing which hours I work.*"

"*I work as I wish to, within my own sense of professionalism and ethics.*"

"...*control over decisions.*"

"*I'm not a good rule-follower; I like making my own rules.*"

"...*the creativity — there's no limitations on my ideas.*"

Identity with results:

"...*make things happen.*"

"...*the adventure of starting something and getting it done.*"

"...*building and creating value.*"

"...*the sense of accomplishment.*"

"...*knowing that I've added value.*"

"...*setting my own goals in life, and then reaching them.*"

"*I'm an artist of necessities; I love filling society's needs.*"

"*I want to be a game-changer.*"

There's no reason we can't create these same motivational forces — empowerment, and identity with results — inside large organizations, at every level. This is the goal of the business-within-a-business paradigm.

What It Means to be a Business Within a Business

The business-within-a-business paradigm doesn't mean operating internal service functions as profit centers. It doesn't require chargebacks, where managers actually pay one another for internal services. And certainly it doesn't imply an arm's-length relationship where support staff don't care about the well-being of the enterprise.

It simply means that every manager thinks and acts like an entrepreneur running his/her own little business.

Regardless of the size of the organization, every group should understand that its purpose is to "sell" its products and services to customers (whether or not money changes hands).

As entrepreneurs, staff should define their catalog of products and services, and know whom their customers are. Customers may be peers within the organization, clients throughout the enterprise, or external customers.

Most every business has competition. When speaking about a company in a competitive market, this is obvious. But the same is true of organizations inside enterprises. Internal service providers compete with both outsourcing and decentralization.

Even if it seems to internal service providers as if clients must work through them, the surest way to lose a monopoly is to behave as a monopolist. Everyone should strive to *earn customers' business through performance,* as the supplier-of-choice in a market that has a right to go elsewhere (even if they really can't).

There are other terms for the business-within-a-business paradigm, such as shared services and "intrapreneurship." [23] They all imply the same thing: empowered, entrepreneurial organizations.

The business-within-a-business paradigm brings out the best in people:

- **Customer focus:** "I understand those are my *customers,* not unruly children or helpless victims of my decisions. In many cases, they have the right to choose what they buy from me."

- **Results orientation:** "It's my product line. I know I'm accountable for delivering everything I promise."

- **Quality:** "I'm proud of my work. It's my job to make it the best."

- **Efficiency and cost control:** "I have to be the best deal in town."

- **Teamwork:** "To stay competitive, I have to focus on my own domain. I get help from peers when I need other specialties."

- **Judicious risk-taking:** "I have no choice but to take some risks to keep up with my competition; but I do so thoughtfully."

- **Use of vendors:** "I manage a business, not just the resources I've been given. If buying something is more cost-effective than building it internally, or if I need more capacity, I'll be the first to offer the 'buy' option alongside the 'make' approach." [24]

- **Innovation:** "I've got to stay ahead of (or at least keep up with) my competition. So I'd better innovate." [25]

These are all traits of successful entrepreneurs.

By the way, here are some things you won't hear from a business-within-a-business organization (all quotes I've actually heard):

- "My job is to get budget at the beginning of the year, and make sure it's used up by year end."
- "We're in public service; we're not a business. So I don't have to listen to any customers."
- "My customer is the enterprise as a whole. I know what's best for you."

Why Not "Partners"

Some people oppose the business-within-a-business paradigm because they believe that the best way to encourage collaboration is to declare groups "partners" in a shared goal.

For example, they'd say that IT and the clients they serve are "partners" in pursuing technology-enabled business strategies. As such, they share authorities and accountabilities.

This may sound nice on the surface. But what does it really mean?

In fact, some interpretations of partnership induce behaviors that *undermine* relationships, not strengthen them. Using IT as the example, let's explore two dangerous definitions of "partners."

Dangerous Definition 1: We share everything. Some say that partnership means that IT staff and clients are one team, and decide everything jointly.

This "all for one, one for all" notion of partnership sounds like it should induce great collaboration. But in practice, respective authorities and accountabilities are unclear, and each party has the right to meddle in the other's domain of expertise:

- IT staff feel they have a right to tell clients how to run their businesses, and may even claim the authority to force changes on clients to make best use of their technologies.

- Clients feel they have the right to tell IT staff how to manage their technology projects.

Instead of each contributing his/her unique competencies, decisions are influenced by people who aren't fully qualified to make them.

Worse, shared accountability is equivalent to *no* accountability. Without clearly defined individual accountabilities, team members struggle for control and projects mire. And when things go wrong, everybody takes cover under the banner of "partnership."

Dangerous Definition 2: We're the experts. Another misguided interpretation sounds something like this: "We are partners, and hence equals. And since we're the IT experts in this partnership, we'll decide what technologies you get."

Note that this is the opposite of customer focus. This "we know what's best for you" attitude can only serve to erode relationships.

Furthermore, this definition of partnership is fundamentally disempowering and unproductive. It would be unfair to hold clients accountable for their business results if they can't control their means of production, including IT. That's the Golden Rule.

Using the concept of partnership to justify disempowering clients leads to resentment, disputes over authority, and strained relationships. It can also induce delays or inaction as the two parties struggle to come to agreement.

A Better Form of Partnership: Customer-supplier Relationships. Wouldn't it be great if, as on Star Trek, we could do a "mind meld" and merge clients' knowledge of their businesses with IT staff's knowledge of technologies!?

But of course we can't. Therefore, one party must share what it knows, and the other party must make the decision.

Dangerous Definition 1, with everybody sharing and everybody deciding, doesn't work. Dangerous Definition 2, where clients share their business knowledge and IT staff make all technology decisions, doesn't work either.

A better answer is found in the business-within-a-business paradigm. Effective partnerships are built on *customer-supplier relationships:*

- Suppliers (like IT) respect customers' rights to make purchase decisions. They present options and share what they know, then let customers decide what they'll buy.

- Customers decide what they'll buy, then let suppliers figure out how to produce it. Suppliers choose their own methods and tools, and manage their staff (including contractors) and vendors. They're empowered and proactive, without disempowering their customers in any way.

Of course, with authority comes accountability:

- Customers have the authority to decide what they'll buy, and hence they're accountable for justifying their utilization of internal services, paying all life-cycle costs, and realizing the benefits.

- Suppliers have the authority to decide how they'll produce those results. And they're accountable for the reliable delivery of products and services at competitive costs.

Customer-supplier relationships are clean and mutually respectful. They match authorities and accountabilities.

Beyond that, they're synergistic. Synergies only result from taking advantage of people's different strengths, and assigning distinct authorities and accountabilities, without any loss of commitment to one another's success. Mutually respectful customer-supplier relationships do exactly that.

Proactive Entrepreneurship

Entrepreneurs strive to please customers. But this does *not* mean that they're passive order-takers. They're proactive in many ways:

As a business within a business, internal entrepreneurs *market the value of their products and services*. This is not meant to be self-serving. It's to lift customers' awareness of the possibilities so as to engender more creative uses of their offerings.

Entrepreneurial organizations *proactively schedule time to talk to customers* about their business challenges. This is "sales" in the best sense of the profession — not pushing products, but helping customers solve their problems and achieve their goals. The result is opportunities closely aligned with customers' objectives.

In response to customers' requirements, suppliers *proactively offer alternative solutions* — as in Chevrolet, BMW, and Rolls-Royce. This way, they make customers aware of better ways to address their needs without going so far as to choose for them.

Internal entrepreneurs *help customers analyze these alternatives* in

Chapter 9: Principle 7: Business Within a Business 93

the context of customers' values (not their own). This isn't a matter of making a recommendation (a form of "we know what's best for you" that takes on some accountability for customers' success). Rather, it's a consultative process that says, "If speed is most important to you, pick alternative A; but if costs are more important, select B."

By the way, if suppliers have done a good job of sharing all they know, it's likely that customers will come to the same conclusions as they did. But if customers do select an alternative other than the one the supplier would recommend, perhaps they know something about their business that suppliers don't know; or perhaps they're applying their own values to the trade-offs. In any case, entrepreneurs respect customers' purchase decisions.

Entrepreneurial organizations can also be *proactive about facilitating enterprisewide decisions* such as policies and standards. Of course, these decisions are made by the community of relevant stakeholders, not unilaterally by any one expert. But suppliers can put forward the issues and coordinate stakeholders' decisions on behalf of the enterprise.

Internal entrepreneurs *proactively invest in their own businesses* — in process improvements, technology innovation, and new products — to remain competitive. They decide their own business strategies, and their research and professional-development priorities.

Entrepreneurs *proactively maintain and evolve their infrastructure.* ("Infrastructure" means assets owned by an organization for the purpose of producing its services.) Internal entrepreneurs don't ask for customers' permission before they buy new manufacturing capacity; they're empowered to acquire whatever they need to satisfy customers' demands for services. Their decisions are based on market needs in total (enterprise capacity plans), not a demand from any single customer.

Entrepreneurs don't wait for customers to tell them to offer new products and services. They *proactively evolve their product lines* (without forcing solutions on customers). They put new products "on the shelf" (making them available to customers), but only take them "off the shelf" (actually deploying them) when customers have agreed to buy them.

And entrepreneurs *proactively reduce costs and improve quality* to remain competitive.

In the business-within-a-business paradigm, everybody is a "product manager." Some sell to external customers, and generate profits for the enterprise. Others sell their products and services internally, probably at breakeven. But all are entrepreneurs who are accountable for managing their businesses.

As proactive as they are, there's a line they don't cross. Entrepreneurs respect that customers know their businesses best, and are accountable for their business results. Therefore, by the Golden Rule, customers must have the authority to decide what they "buy" from internal service providers.

Implications for Structure

The business-within-a-business paradigm sends all the right signals and rewards the right behaviors. It harnesses everybody's creativity. It aligns the work of customers and suppliers. And it builds highly effective partnerships.

Organizational structure is the key to achieving such an entrepreneurial organization. The guideline for structure is: **Define groups as lines of business, bounded by the products and services they produce** — by what each "sells," not what it does. (Again, I'm not suggesting that money changes hands.)

Designing a healthy structure is a matter of dividing the organization's various internal and external products and services among its groups, laying out a mosaic of entrepreneurships.

This capstone Principle creates empowered jobs, where staff are customer focused, accountable for results, creative and entrepreneurial, and highly motivated.

SYNOPSIS

» In a healthy organization, everybody thinks and acts like entrepreneurs running small businesses within a business.

» Internal entrepreneurship is highly motivational and productive.

» The best partnerships are built by internal customer-supplier relationships (not shared accountabilities and authorities).

» Define groups as lines of business, bounded by the products and services they produce.

~ PART 3 ~
The Building Blocks of Organization Charts

The beginning of wisdom is the definition of terms.

Socrates

An engineering science comprises both principles and components. Part 2 described the seven Principles of structure. This Part describes the components.

Designing a healthy structure certainly is not child's play. Nonetheless, it's useful to think of these components as "Building Blocks" that you can assemble into an organization chart.

The way the Building Blocks are defined is crucial. The Principles guide us in how to do that (as well as in how to assemble them).

The most salient to the definition of the Building Blocks is Principle 7, the business-within-a-business paradigm. It tells us that the Building Blocks should be the *lines of business* that exist within organizations. Then, when you assemble the Building Blocks into an organization chart, you're assured that every job is an empowered entrepreneurship.

This Part defines all the lines of business that exist within organizations.

Chapter 10:
Overview of the Building Blocks

The **Building Blocks are the lines of business that exist within organizations.**

They appear in every industry — corporations, not-for-profit organizations, higher education, governments, and even clubs.

These same lines of business exist within departments inside enterprises. Like whole companies, most internal service providers (such as IT, HR, Finance, and Facilities) include operations functions that produce ongoing services (like manufacturing), gurus in disciplines (like engineering), customer service, and even internal sales and marketing (the relationship managers that link an internal service provider to the rest of the enterprise).

Some departments may not include all the Building Blocks. Nonetheless, their work can be defined in this same language.

The Building Blocks help you avoid gaps (Principle 3). They're essentially a checklist of all the possible lines of business in any organization, so that you can be sure that you have all the pieces you need somewhere in your new organization chart.

All the Building Blocks are important. Some may be larger than others. Some may be more strategic to future growth. Some may serve clients, while others serve customers within the organization itself (internal support services).

But they're all businesses, and all should be creative, customer focused, and entrepreneurial. They're all essential, and should be treated with respect. Any biases or discrimination can only serve

Chapter 10: Overview of the Building Blocks 99

to diminish the effectiveness of the victimized group, and ultimately the performance of the entire organization. In a healthy organization, **there are no second-class citizens**.

At a high level, there are five Building Blocks — five types of businesses. Each has sub-categories. (Figure 3 lists the high-level Building Blocks.) The Chapters in this Part define each of these five, and the more detailed lines of business within them.

Figure 3: The Building Blocks of Structure

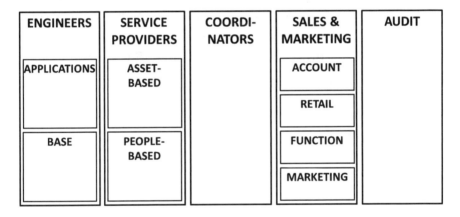

Before we explore each of the Building Blocks, I must give you two cautions:

Caution 1: This framework of Building Blocks is **not a prescribed organization chart.** There's no such thing as an off-the-shelf organization chart that's right for everybody.

Rather, the Building Blocks are simply a **language you can use to describe organization charts**.

Too often, leaders get together to plan a new organization chart;

but each leaves the meeting with a somewhat different understanding of words like "engineering" and "operations." The Building Blocks provide a clear, precise, common language for analyzing, discussing, and designing organization charts.

Executives can use this common language to diagnose their current structure; understand one another's proposals; make structural decisions in a factual, analytic manner; and establish a common understanding of their missions and their boundaries.

Caution 2: Some of the names of these Building Blocks may sound familiar. For example, you may have an "Engineering" department today. But it may, in fact, be a combination of multiple Building Blocks; and pieces of the "Engineers" Building Block may be found in other departments.

So please **don't confuse the name of a Building Block with the name of a group in your current organization.**

SYNOPSIS

» The "Building Blocks" are a framework of all the lines of business within organizations.

» The Building Blocks provide a language for describing organization charts (not a recommended structure).

» Don't confuse the name of a Building Block with the name of a group in your current organization.

Chapter 11:
Distinguishing Engineers and Service Providers

By "line of business," I don't mean industry. Who's in the business of, say, airplanes? Boeing, all the airlines, the companies that clean the planes, the companies that cater the food, the local government organizations that run airports, the federal government that controls air traffic and sets the rules, and more.

You can't just say "airplanes" and know what business people are in. Similarly, within an enterprise, you can't just say "operations" or "infrastructure" and know what business staff are in. The framework of Building Blocks has to be more refined.

Let's begin to tease apart the lines of business within organizations by considering the difference between Boeing (which sells airplanes) and an airline (which sells a transportation service). Or consider Ford and Hertz; or a building contractor and a hotel.

Pick any industry, and you'll see two very different kinds of businesses:

- **Engineers** design and produce products (assets), and support those assets with their design expertise.

- **Service Providers** buy those assets, own and operate them, and use them to sell services.

For most any type of asset, you'll see both. For example, with regard to airplanes, Boeing is an Engineer, and an airline is a Service Provider. For cars, Ford is an Engineer; and Hertz is a Service Provider.

The same split occurs in internal service providers. In IT, for example, there are Engineers who implement computers, storage devices, networks, and applications. And there are Service Providers who own and operate those assets to deliver services such as computer time, data storage, connectivity, applications hosting, and software as a service (e.g., email).

Engineers and Service Providers are very different businesses — differing in their competencies, business models, cultures, and their products and services. Thus, they are separate Building Blocks.

Of the five categories of Building Blocks, Engineers and Services Providers are the most familiar. So we'll begin with a Chapter on each of them.

SYNOPSIS

» Although they may work with the same things, Engineers and Service Providers are very different. They are two distinct Building Blocks.

» For almost any type of asset, you'll see both. For example, with regard to airplanes, Boeing is an Engineer, and an airline is a Service Provider.

Chapter 12: Engineers

Engineers create the organization's products. [26] They are the discipline or technology specialists, the designers, and the gurus in those products.

Engineers maintain a locus of expertise in a specific engineering discipline, technology, or professional specialty. They use their expertise to **design, build, and install "solutions,"** perhaps utilizing vendors' products and then adding value. A solution may be a physical asset, software, or a design of intellectual property.

They also **enhance, tune, repair, configure, and support those solutions**. This definition does not distinguish those who design and build products from those who repair them. Essentially the same bottom-of-the-T is required to do both. Engineers sell anything that requires in-depth expertise in the design of the organization's products — knowledge of what's inside that "black box."

Here are some of the services of Engineers, all based on their in-depth product-design expertise:

- Solutions (entirely new, or enhancements)
- Repairs
- Configuration tuning
- Domain specific incident management (a.k.a., second-level support, as a point of escalation for customer-service)
- Documentation and training materials, and training

- Expert time, studies, presentations, and sales support

Engineers can be found in any industry. Here are a few internal support functions and industries, and examples within each:

- **IT:** applications developers, infrastructure systems engineers
- **HR:** compensation and benefits design, performance-management systems design
- **Finance:** tax law, investment financial analysis
- **Product manufacturers:** product designers, manufacturing engineers
- **Health care:** outcomes management, doctors
- **Education:** schools and professors, curriculum and instructional design
- **Government:** policy and program design, decisions on grants

Engineers don't own the solutions they produce (other than work in progress). They sell solutions to others, be they peers within the organization (such as Service Providers) or external clients.

Applications versus Base

In many functions and industries, there are multiple layers of Engineers. Some design fully assembled products; others are experts in components that go into various products.

In automobile engineering, for example, there are those who design cars, and those who design trucks. Both of these "applications" draw on "base" components like engines, transmissions, and electrical systems.

Figure 4: Applications versus Base Engineers

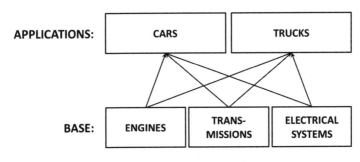

In some fields, there are many layers of engineers. In IT, for example, there are computer engineers, databases and middleware, and applications. Each layer draws on the lower layers, and supports the next layer up.

Those who assemble components into complete products (the top layer) are called "**Applications Engineers.**" They produce **purpose-specific products** which are tailored to various kinds of customers' needs.

All lower layers are termed "**Base Engineers.**" They design the components that go into various applications. These components are **purpose-independent** in that they serve multiple applications.

So, engineers who build cars and those who build trucks are Applications Engineers; they specialize in different purposes. Base Engineers contribute parts and advice to both.

At Boeing, Applications Engineers design different kinds of airplanes. They employ Base Engineers who design engines, control systems, interiors, etc.

In higher education, the various schools are Applications Engineers; curriculum and instructional design experts are Base Engineers.

In IT, the term "application" refers to data-object-specific software — systems designed to handle information about a particular topic. Base Engineers sell and support data-object-independent technologies such as computing platforms, database engines, middleware, and telecommunications networks (infrastructure), as well as end-user-computing and data-analysis tools, software-engineering tools and methods, and models such as artificial intelligence.

Bottom of the T

For Engineers, the bottom-of-the-T is a specific domain of technology, field of science, professional discipline, or branch of engineering.

Of course, they have to deliver practical solutions. So a significant top-of-the-T competency is engineering methods and tools (e.g., for design and testing) as well as project management.

If they're expected to produce anything other than solutions, their depth in their professional domain is bound to suffer.

For example, if they're not given clear requirements for the things they're to design, they may have to learn about customers' businesses to help customers define those requirements (the product of a different Building Block). Any time they spend studying customers' businesses is time away from their real specialty.

Figure 5 summarizes the core competencies of Engineers.

Figure 5: Competencies of Engineers

Bottom of the T: A domain of technology, science, or a discipline
Key methods: Design and testing, project management
Technical skill: High
Project management: High
Service management (operations): Low
Business of organization's customers: Low
Interpersonal skills: Low
People supervision: Medium

To avoid conflicts of interests (as per Principle 5), it's important to know the biases of each Building Block.

Engineers love innovation, but they quickly become bored (and may get sloppy) when asked to do routine "keep it running" work.

They're laser focused on their domain of expertise, and love to study its nuances and leading-edge discoveries. They count on other equally focused Engineers when other skills are needed.

Engineers are willing to produce anything their customers need — from enterprisewide to highly customized solutions. But they're not experts in customers' businesses. And they cannot assess customers' needs in an unbiased, business-driven manner, since they're paid to live and breathe their own domain of solutions.

Considering the five fundamental conflicts of interests in organizations described in Chapter 7 (Principle 5), Engineers' biases are summarized in Figure 6.

Figure 6: Biases of Engineers

- **Invention versus operational stability:** Invention
- **Purpose-specific solutions versus components:**
  Applications: purpose-specific
  Base: components
- **Enterprisewide thinking versus the focus of a specialist:** Specialist
- **Technical specialization versus unbiased diagnosis of clients' needs:** Technical specialization
- **Service oriented versus audit:** Service oriented

SYNOPSIS

» Engineers are experts in technologies or disciplines.

» Engineers design, build, sell, repair, and support solutions (products).

» There are two types of Engineers: Applications (purpose-specific solutions) and Base (purpose independent, e.g., component parts).

Chapter 13:
Service Providers

Service Providers deliver ongoing services. [27] They may buy products from Engineers, and then use those assets to deliver a service (like an airline that buys airplanes and sells a transportation service). Or the service may be delivered primarily by people.

The distinguishing attribute of Service Providers is the ongoing nature of services. This doesn't include a stream of small, unique projects (like the repairs that Engineers sell). Rather, it's essentially the same routine service, day after day. Think of services as water flowing from a spigot.

Service Providers keep things running reliably, safely, and efficiently.

This Building Block includes manufacturing, customer service, and a range of support services.

Asset-based

Some **services are based on the ownership of assets,** where their customers are essentially buying use of those assets. These are termed "Asset-based Service Providers." [28]

Their business model is to look for cases where customers can share the use of an asset. They acquire those assets, termed "infrastructure," from the organization's Engineers (or directly from vendors), and use them to produce services.

In IT, for example, we may all own our own PCs; but it's not possible for each of us to run our own networks and data centers.

Asset-based Service Providers see this as a business opportunity; they acquire networks and shared-use computers, and sell the use of them to others.

Assets include information as well as tangible properties.

Some examples of <u>Asset-based</u> Service Providers include:

- **IT:** computer time, data storage, and connectivity
- **HR:** benefits administration, employee data administration
- **Finance:** accounting services, treasury
- **Product manufacturers:** factories, warehouses, logistics
- **Health care:** hospital rooms, clinics, labs
- **Education:** classrooms, residence halls, library
- **Government:** roads, parks, ports and airports, air traffic control (the air space), welfare, data services

Asset-based Service Providers are not technology specialists. They rely on Engineers to build, document, upgrade, and repair their assets. They know how to operate their infrastructure; but their real expertise is in providing services, and all that entails.

People-based

People-based Service Providers sell **services produced by people** rather than assets. Equipment, such as computers, is employed only to make the people more productive at tasks they conceivably could do manually. But the use of the asset is not really what customers are buying; they're buying "use" of the people.

Some examples of <u>People-based</u> Service Providers include:

- **IT:** service desk, field technicians, writers and trainers
- **HR:** counseling, recruiting, onboarding, job grading
- **Finance:** procurement, customer service
- **Product manufacturers:** customer service, field technicians
- **Health care:** admissions, nursing, non-medical services
- **Education:** registrar, admissions processing, student counseling, business services
- **Government:** social services, elections, weather forecasts

There are three types of People-based Service Providers:

- **Product support:** These services help customers get value from the rest of the organization's products and services. A common example is customer service (a help desk). Another example is training in use of the organization's products.

- **Internal support:** These services leverage the time and enhance the abilities of others within the organization.

  An example is a project management office (PMO) which helps others manage their projects, offering services such as project plans (PERT charts, work-breakdown structures) and project tracking and reporting. (They don't manage projects. The problems with a PMO that's accountable for projects are described in Chapters 5 and 37.)

  Other examples are field technicians who follow the instructions of other Building Blocks, and technical writers who help others communicate their knowledge.

- **Business services:** These services are outside the primary intent of the organization.

  For example, Administration, IT, Finance, and HR departments contain all the Building Blocks. But from a CEO's perspective, they're all classed as People-Based Service Providers.

  Similarly, when internal services are decentralized (such as a finance or administration function within an IT department), they're classed as People-based Service Providers.

Figure 7: Three Types of People-based Service Providers

ENGINEERS	SERVICE PROVIDERS	COORDI-NATORS	SALES & MARKETING	AUDIT
APPLICATIONS	ASSET-BASED		ACCOUNT	
BASE	PEOPLE-BASED		MARKETING	

- Product support
- Internal support
- Business services

Bottom of the T

The specialized expertise of both types of Service Providers is *service management*. This includes assessing the market and determining which services to offer; defining those services; acquiring and managing any needed assets, methods, and vendor services; contracting with customers (service-level agreements);

Chapter 13: Service Providers 113

managing capacity; and managing the ongoing delivery of those services.

Figure 8: Competencies of Service Providers

Bottom of the T: Specific services and how they're produced
Key methods: Service delivery and management
Technical skill: Medium
Project management: Low
Service management (operations): High
Business of organization's customers: Low
Interpersonal skills: Asset-based: Low
  People-based: High
People supervision: Asset-based: Medium
  People-based: High

When it comes to their biases, Service Providers are responsible for stability, responsiveness, reliability, security, and low cost. They readily solve problems in service delivery, continually improve existing services, and offer new services when the time is right.

But major innovations (inventions) inevitably disrupt smooth operations. Service Providers are an intentional damper on innovation, proceeding only when potential market demand justifies the investment and the technologies have evolved to the point of being ready to produce reliable, safe services. They're biased to favor stability over innovation; they change only when it's safe.

Service Providers' biases are described in Figure 9.

Figure 9: Biases of Service Providers

- **Invention versus operational stability:** Operational stability
- **Purpose-specific solutions versus components:** Agnostic
- **Enterprisewide thinking versus the focus of a specialist:** Specialist
- **Technical specialization versus unbiased diagnosis of clients' needs:** Unbiased
- **Service oriented versus audit:** Service oriented

SYNOPSIS

» Service Providers produce an ongoing stream of services. They keep things running reliably, efficiently, and safely.

» Asset-based Service Providers own infrastructure, and sell services based on those assets.

» People-based Service Providers sell services produced primarily by people.

Chapter 14: Coordinators

Some things require a consensus among stakeholders — people throughout the organization, and in some cases clients as well. The next Building Block drives those shared decisions.

Take, for example, planning. Every internal entrepreneur is responsible for planning his/her own business within a business. That's empowerment. But all their plans must be coordinated to add up to the organization's plan.

Furthermore, planning is a specialty in its own right; there are methods and analytical skills which require study.

Thus, there's a role for a "Planning Coordinator" whose job is not to decide the plan, but rather to help everybody develop their own plans *in a coordinated manner*. This example illustrates the "Coordinators" Building Block.

Coordinators help stakeholders (within the organization and beyond) come to agreement on shared decisions such as policies, plans, and standards.

They establish processes for decision making. They enable those processes with methods and project plans. They ensure that the right stakeholders are involved. They provide common information, such as assumptions and trends, templates and formats, conceptual frameworks, and timeframes. They help individuals with their respective duties in the process. And they facilitate collaboration on interdependencies, such that each group's plan fits into a higher-order enterprisewide or organizationwide plan.

(You're not a Coordinator if, like all the other stakeholders, you participate in these decisions. You're only a Coordinator if your job is to facilitate the decision process.)

Coordinators also help people utilize the resulting shared decisions. For example, they compile the results and make them available. And they help others find, interpret, and apply the relevant policies, standards, and plans.

But Coordinators are not accountable for the content; the participating stakeholders are. Coordinators are just accountable for effective consensus-based decision-making processes.

And if stakeholders can't reach consensus, there's little a Coordinator can do. It would be a mistake to give Coordinators the authority to force a decision. Since they're not accountable for others' results, doing so would violate the Golden Rule. Perhaps it's premature to expect a consensus. But if it's important to make a decision now, Coordinators can help executives motivate stakeholders to want to agree.

Types of Coordinators

There are numerous topics which require coordination. Each is a distinct line of business. Some are business oriented, while others are more technical and product specific. Most are common to all functions; a few are function-specific.

Common business-oriented Coordinators include the following:

- **Strategy Planning:** A strategic plan answers the questions, "What businesses do we want to be in? And how will we get from here to there?" The answers are based on a continually changing environment, and the organization's strengths, weaknesses, opportunities, and threats.

Entrepreneurs at every level of an organization are accountable for the strategies of their own groups. But their individual strategies must fit together to achieve the strategies at the next level up.

And strategy planning is a science in its own right. So leaders need help with the process as well as the coordination.

A Strategy Planning Coordinator is an expert in strategic planning methods, and in the business environment (including competitors) and how that environment creates threats and opportunities. This function doesn't decide strategies, but rather helps everybody decide their own respective strategies, in a coordinated fashion.

- **Operational Planning:** An operating plan looks one or two years ahead, and answers the questions, "What do we plan to deliver in the coming year? And how will we fulfill that demand (including the needed resources, such as budgets)?"

  An Operational Planning Coordinator is an expert in business and budget planning methods (especially investment-based budgeting [29]), and in the tools used to create an operating plan and budget based on forecasted demand.

- **Research:** Within organizations, research means investigating new products, services, technologies, or disciplines. In a manufacturing company, research may develop new products or manufacturing techniques. In IT, an example is the investigation of new vendor products and services.

A Research Coordinator doesn't *do* any research. Rather, he/she helps others do research, by sharing expertise in research methods and in how to develop research proposals. He/she also helps the executive decide which research projects to fund, as a portfolio of investments aligned with the organization's strategies.

This function is akin to the US National Science Foundation which distributes research grants to universities and corporations (who do the actual research), but doesn't do any research itself.

- **Organizational Effectiveness:** The focus of this Coordinator is on *how* the organization does business (not *what* business it produces). His/her scope includes: culture, structure, resource-governance processes (the "internal economy"), shared methods and tools, and metrics and rewards. [30]

    Organizational Effectiveness is distinct from Human Resources in that it engineers the "organizational ecosystem," whereas HR focuses on the employment relationship.

    An Organizational Effectiveness Coordinator is an expert in the principles of design of those organizational systems, and in the methods of change (not just generic change management, but methods specific to engineering organizational system).

    This Coordinator leads transformation projects (including a restructuring), and then helps everybody work within the organizational design. An Organizational Effectiveness Coordinator may also coordinate employee communications.

- **Audit Response:** Coordinators do not audit or judge the people they're meant to serve. But external parties may audit

Chapter 14: Coordinators

the organization; and when that happens, the organization must provide a point of contact, and a coordinated response.

An Audit Response Coordinator helps everybody respond to auditors and provides a point of contact, but the content of the audit response is the responsibility of the appropriate groups throughout the organization.

Response includes the initial provision of information, as well as coordinated projects to remediate any problems revealed by audits.

With his/her knowledge of the questions auditors ask, an Audit Response Coordinator can also offer "assessments" to help others prepare for an audit. Assessments are a voluntary service, not an audit; results are provided strictly to those being inspected.

- **Regulatory Compliance:** Everybody is accountable for complying with laws and regulations. A Regulatory Compliance Coordinator helps them do so by providing expertise in laws and regulations, and in how they apply to the organization. (See Chapter 38.)

  He/she also provides a point of contact for regulators, and coordinates any compliance examinations and remediation projects.

  And like the Audit Response Coordinator, he/she may offer internal assessments to help others comply.

- **Business Continuity:** In case of a disaster, the organization must ensure that its staff are safe, stabilize the situation, and then bring the business back to normal operations (disaster recovery). Beyond that, business-continuity planning can mitigate the damage done by disasters.

A Business Continuity Coordinator helps everybody develop their own plans, and coordinates interdependencies to optimize the resilience of the organization. This function also coordinates tests of the plan, and (if triggered by an actual incident) the execution of the plan.

- **Security:** Security means minimizing the risks of harm from espionage, theft, and sabotage.

In an empowered organization, everybody is responsible for their own security, and for providing safe, secure products and services to their customers. A Security Coordinator helps them do so.

A Security Coordinator is an expert in threats. This function helps everybody come to consensus on security policies; keeps people informed of threats and defense strategies; leads investigations of and responses to security incidents; and coordinates recovery and mitigation projects done by the appropriate managers.

Again, this is not an audit function. A Security Coordinator does not check up on everybody's compliance with policies. And he/she is not a Service Provider offering ongoing services like guards, building access control, or IT firewalls and identity management. A Security Coordinator works at a higher, organizationwide level, as a service to others who are accountable for their own safety.

In IT, this is titled the Chief Information Security Officer (CISO).

Chapter 14: Coordinators 121

Common <u>technical</u> Coordinators include the following:

- **Standards:** Organizations set product design standards to ensure the interoperability and supportability of their products.

    In your home, the shape of the electrical outlet is an example of a standard. In manufacturing industries, a simple example is constraints on the variety of nuts and bolts used in products. In IT, standards are protocols, APIs, and interfaces.

    Standards are conscious constraints on design (not preferred brands and models). They permit evolutionary change without loss of integration and interoperability.

    There are many stakeholders who are affected by standards: Engineers, Asset-based Service Providers, some People-based Service Providers, and in some instances, clients. All their perspectives should be considered before a standard is decided. Not only is their knowledge valuable; their involvement in the decision encourages compliance.

    Therefore, a Standards Coordinator does not decide standards. Rather, he/she creates a framework. Then, on an ongoing basis, he/she pulls together the appropriate stakeholders to decide each "cell" within the framework (specific standards), and helps them come to consensus.

    Also, a Standards Coordinator helps Engineers access and interpret agreed standards in the course of designing solutions.

- **Design Patterns:** To optimize the best interests of customers who buy multiple things from the organization, the various products and services should be designed to fit well together. This involves agreements among the various Engineers and Service Providers about "design patterns."

A Design Patterns Coordinator is an expert in the "ripples" — how decisions about, or changes in, one domain affect others. He/she facilitates consensus on design guidelines that affect all domains, and helps individuals apply those guidelines to their decisions and understand the ripples.

In your community, city planning and zoning is an example, dictating which types of buildings belong in each area of the city.

In a distribution company, the pricing, sales incentives, and marketing promotions for one product line may affect the company's ability to sell other product lines. Impacts cross channels, product categories, and services (such as distribution, sales, and marketing). Coordination of policies, and of requests for variances, is needed to reduce the risk of optimizing one product while sabotaging the sales of others.

In higher education, a Design Patterns Coordinator drives enterprise programs to enhance student success, and helps multiple groups package offerings for different audiences (e.g., for non-traditional students).

In IT, design patterns suggest where various functions fit in a broader enterprise architecture, e.g., what software goes on personal devices (PCs and mobile devices) versus on shared servers. And he/she maintains a map of how existing systems are interconnected and how data flows through them, to help Engineers anticipate how changes in one system will affect other systems.

In IT, the combination of Standards and Design Patterns Coordinators is called an "enterprise architect."

Some function-specific Coordinators are:

- **Information Policy:** In IT, there's a need for coordination of policies with regard to how information is handled. One example is retention policies (particularly complex given that electronic records are interdependent, so information that has to be retained may depend on the availability of other data that should be destroyed).

    Other examples include how the system-of-record is determined; how data owners are chosen, and their accountabilities and authorities; what information is to be shared (i.e., access-rights); how privacy is maintained; and what may and may not be said using enterprise resources such as web sites, blogs, and electronic mail.

- **Employment Policy:** In HR, employees are hired with the expectation that they'll evolve in their careers through multiple jobs within the enterprise. Staff are considered employees of the enterprise, not just the group that initially hired them. Thus, some decisions about the employment relationship should be made by consensus, not by each department alone (or by HR).

    An example is pay. Compensation must be coordinated to avoid inequities, so that two people doing similar jobs in different places in the enterprise are paid roughly the same. An enterprise may collectively decide to pay above market to attract top talent, or below market to improve near-term margins. This is a business decision, not strictly an HR decision.

    The role of an Employment Policy Coordinator is to bring enterprise stakeholders to consensus on such policies.

Bottom of the T

All the Coordinators are similar in that they facilitate and coordinate shared decisions. They're experts in identifying stakeholders and in bringing teams to consensus.

In addition, each is an expert in the *structure* of the content they are coordinating. They know enough about that content to understand the trade-offs in decisions; but they cannot be expert in everybody's domains to the point of deciding that content, and they must not disempower others (even if they do know a lot about the subject).

For example, a Planning Coordinator knows what goes into a good plan. But the actual content of the plan comes from the participating managers.

Figure 10 summarizes the core competencies of Coordinators.

Figure 10: Competencies of Coordinators

Bottom of the T: The topic being coordinated
Key methods: Consensus building, information structuring
Technical skill: Low (business) to Medium (technical)
Project management: Medium
Service management (operations): Low
Business of organization's customers: Low
Interpersonal skills: High
People supervision: Low

Chapter 14: Coordinators

It's critical that Coordinators are unbiased facilitators of shared decisions. But they do have biases, as described in Figure 11.

Figure 11: Biases of Coordinators

- **Invention versus operational stability:** Invention
- **Purpose-specific solutions versus components:** Agnostic
- **Enterprisewide thinking versus the focus of a specialist:** Enterprisewide (or organizationwide) decisions
- **Technical specialization versus unbiased diagnosis of clients' needs:** Unbiased
- **Service oriented versus audit:** Service oriented

SYNOPSIS

» Coordinators help others, in part by helping them come to consensus on shared decisions such as plans, policies, and standards.

» Business-oriented Coordinators include business planning, organizational effectiveness, audit response, compliance, business continuity, and security.

» Technical Coordinators includes standards and design patterns.

» Functions may have specific types of Coordinators, like information policy in IT and employment policy in HR.

Chapter 15:
Sales and Marketing

The fourth of the five Building Blocks, Sales and Marketing, is the client-facing part of an organization. [31] It adds value in two ways:

- It **enhances the organization's relationships with clients.**

    This is not a single point of contact, but rather a default point of contact. Sales facilitates effective communications between clients and everybody in the organization.

- It helps clients address their challenges (problems and opportunities) using the organization's products and services.

    Sales staff help clients acquire just the right products and services from the organization's entire catalog. In this way, they **align the organization with clients' strategies and business needs,** and **help clients discover high-payoff opportunities for the organization's products and services.**

Caution: The word "sell" has two meanings: to provide a product, or to help customers buy others' products. This Building Block does the latter. It *brokers* sales; but it's not accountable for the delivery of other groups' products and services.

Internal Service Providers Need Sales Too

In companies, the need for Sales and Marketing is obvious. But this Building Block is equally essential to internal service providers. They have tough competition from decentralization and outsourcing. And despite monopolies in a few areas, they must earn "market share" by developing great relationships with clients.

Furthermore, an internal service provider must find ways to contribute value to clients' critical challenges and strategic opportunities. Sales provides the linkage to clients and expertise to do that.

In companies, the goal of Sales is to maximize revenues. But for internal service providers, the goal is not to convince clients to spend more (an expense to the enterprise). Rather, it's to maximize the organization's contributions to clients' success, i.e., to generate the *right* business that delivers the most value.

Internal service providers may not want to call this function "sales." The term may offend some clients (because they don't understand its value). And you wouldn't want to imply that you're out to get as much money as possible from internal clients. Within internal service providers, Sales may be named "business relationship managers" or "consultancy." (The "Chief Digital Officer" role, in its best incarnations, is a Sales role within IT. [32])

What Good Selling Is Really About

External or internal, and whatever the name, great salesmanship focuses on helping clients, not on helping the organization "push" its products and services. In the sales literature, this is termed strategic selling, consultative selling, or partnership selling.

Great relationships are built by recognizing what's unique about each client, understanding his/her business challenges, and connecting him/her with the right suppliers in the organization. That's why Sales staff spend most of their time with clients.

Beyond just relationship building, Sales helps clients discover creative, high-payoff uses of the organization's products and services. That linkage between clients' business and an organization's deliverables is sometimes called "strategic alignment."

This is not a matter of passive order-taking. Sales proactively meets with key clients to discover ways the organization can help them achieve their goals. But it always works in a business-driven manner, not pushing any particular products or services.

Similarly, the purpose of internal Marketing is not to aggrandize the organization. It's to help clients understand the organization's value, so they'll make better use of, and gain more benefit from, its offerings. Marketing also helps an internal service provider better understand clients' needs through market research.

Decentralized Sales Functions

Some internal service providers decentralize the Sales function, having it report to clients rather than to the shared-services organization.

It's true that Sales staff must be very close to clients, and spend most of their time with clients. They may even be substructured by client business unit (dedicated to specific clients). But that doesn't mean they should report to clients. **Here are some of the problems with decentralization of internal Sales:**

When this function has been decentralized, it seems to devolve into a full-service support function, rather than remain focused on just the Sales Building Block for the shared-services organization.

Also, decentralized Sales staff often seem to treat the centralized department as an adversary, as if their role is to defend their business units against them, rather than facilitate great relationships between the two organizations.

Decentralizing this function makes it hard to balance workloads, for example, where a number of small business units share one Sales person while the large business units may warrant a team.

There are generally difficulties introducing essential methods of Sales for lack of a common boss.

They often miss cross-business-unit opportunities, such as when a "consortium" of clients band together to share an asset such as ERP software.

And decentralized Sales staff generally don't provide services to the other Building Blocks within the shared-services department, like briefings on business trends and client interactions.

Some might believe that they'll have better access to clients if they're part of clients' business units. But in practice, rank often gets in the way. For some reason, it's easier for a shared-services person with a "Director" title to attend meetings of business-unit Vice Presidents, than for a Director within that business unit.

For many reasons, decentralization of the Sales function is not advisable.

Three Types of Sales

There are three distinct types of Sales (not counting Marketing):

"Account Sales" are high-level account representatives who are responsible for *entire client accounts,* regardless of geography. In companies, these might be called "named account managers," where the accounts are one or more global companies. In internal service providers, these might be the "business relationship managers," where the accounts are one or more business units.

Account Sales staff are responsible for the relationship with the entire account. In addition, they proactively meet with selected client executives, and produce a stream of strategic projects by helping these key clients discover high-payoff opportunities.

This is a very senior function. Account Sales professionals can (and should) attend clients' executive-level meetings and credibly discuss business strategies and challenges (without descending into product discussions). They're the kind of people clients would like to hire for senior management positions.

While Account Sales staff may not manage many people or a big budget, their strategic impact is immense and their job grade should be as high as most senior leaders in the organization.

"Retail Sales" staff are not dedicated to any specific accounts; instead, their *territory is geographic*. In companies, this is the geographic sales force and any retail storefronts for walk-in customers. In internal service providers, these are the default points of contact for clients' inquiries and concerns. They may even manage a showroom or demonstration center.

Retails Sales may also include business analysts who convert high-level opportunities (discovered by Account Sales) into detailed requirements.

To differentiate these first two types: Account Sales are proactive, and provide premium service to selected clients (the top executives and key influencers of business strategies). Retail Sales are reactive, and are available to anyone on demand.

Both types of Sales staff are more than just order takers. They add value by helping clients understand what they most need to buy from the organization to address their business challenges.

Figure 12: Three Types of Sales and Marketing

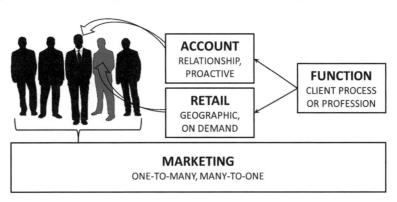

"Function Sales" staff are experts in *clients' professions or specific business processes* — disciplines that are relevant to multiple accounts (hence needed by multiple Account Sales staff).

They don't have a territory. They're a "second-tier" salesforce called in by Account and Retail Sales to help make presentations and diagnose clients' needs.

An example was found in a medical device manufacturer, where Account Sales staff had responsibility for hospitals. When they demonstrated their device in the hospital's *hematology* laboratory, they brought in a trained hematologist. When that same Account Sales representative demonstrated that same device in that same hospital's *immunology* laboratory, they brought in a trained immunologist. That second-tier Function Sales force was shared by all the Account Sales representatives.

In IT, an example might be experts in "digital enterprise" — the application of technology to the enterprise's relationship with its customers. The expertise of a "Chief Digital Officer" can be applied to many business units (hence, it's not an Account Sales function). And it draws on many different technologies (hence, it's not an Engineer function).

Marketing

Marketing focuses on *clients as a whole* — either all, or segments with similar needs and buying patterns. This distinguishes it from Sales which works with clients individually.

Marketing includes two sub-specialties: Marketing Communications (one-to-many), and Market Research (many-to-one).

"Marketing Communications" helps others in the organization communicate with clients. Across all communications channels, Marketing coordinates the organization's messages for consistency with its strategies and its brand, and to ensure that the various entrepreneurs within the organization don't over-saturate the channels.

Marketing Communications includes branding and marketing communications strategies; publications such as brochures, web sites, and newsletters; advertising and direct communications; promotions; and customer events.

In companies, its goal is to generate demand. For internal service providers, it's to encourage understanding of the value of the function and improve client satisfaction.

"Market Research" is the inbound communications channel, serving the organization as a window to the entire client community (its market). Market Research answers questions about what clients think, want, need, and will buy. It ranges from simple customer-satisfaction surveys to complex buying-pattern analyses and market-demand forecasting models.

Bottom of the T

It takes the right people to build an effective Sales and Marketing function.

Great relationships and strategic value both depend on people who know clients, their businesses, and their strategies. Ideally, Sales staff have a background much like clients, with similar education, and experience in the industry and in clients' jobs.

Additionally, Sales staff have a deep understanding of the *linkage* between the organization and clients' businesses. They understand how the organization's products and services can enable clients' strategies.

Of course, Sales staff need excellent interpersonal skills. (There's more on the competencies of Sales in Chapter 46.)

Their bottom-of-the-T also includes methods to identify clients' key concerns and needs; to discover high-payoff, strategic opportunities; and to quantify the benefits of proposed solutions (strategic as well as cost savings).

Marketing is also customer facing. They are experts in communications and market-research methods, and in the way customers think about the organization's products and services.

Figure 13: Competencies of Sales and Marketing

Bottom of the T: Clients, their business strategies, and the linkage to the organization's products/services

Key methods:
Sales: opportunity identification and requirements planning, relationship management, benefits estimation, contract brokerage
Marketing: market communications, marketing strategy, market research

Technical skill: Low ("smart buyer's" knowledge of the organization's entire product line)

Project management: Low

Service management (operations): Low

Business of organization's customers: High

Interpersonal skills: High

People supervision: Low

(An overview of the services of Sales and Marketing is available in the Supplement to this book; contact <ndma@ndma.com>.)

To align with clients' business strategies, an organization must be prepared to flexibly offer any subset of its products and services in a completely business-driven manner. The Sales and Marketing function represents the entire organization without any product bias.

But they do have other biases, described in Figure 14.

Figure 14: Biases of Sales and Marketing

- **Invention versus operational stability:** Agnostic
- **Purpose-specific solutions versus components:** Agnostic
- **Enterprisewide thinking versus the focus of a specialist:**
  Account: Specialist (in clients)
  Others: Enterprisewide
- **Technical specialization versus unbiased diagnosis of clients' needs:** Unbiased!
- **Service oriented versus audit:** Service oriented

SYNOPSIS

» Sales and Marketing are experts in clients, and in the linkage between clients' businesses and the organization's products/services.

» For internal service provides, Sales staff may be called "account representatives" or "business relationship managers."

» Sales is one-to-one, and helps translate clients' business challenges into requirements for the organization's products and services.

» There are three types of Sales: Account (dedicated to specific sets of clients), Retail (geographic, default point of contact), and Function (second-tier sales that's expert in clients' professions or processes).

» Marketing works with the client community as a whole, and includes both marketing communications and market research.

Chapter 16:
Audit

The Audit Building Block includes more than traditional financial and compliance auditors. Any function which **inspects and judges others, and reports the results to someone other than those being inspected,** is considered Audit.

Audit may even have some **authority to block others** (veto their requests or decisions). This is the only Building Block with such power; all the rest focus on serving, not judging, others.

Audit is distinguished from service-oriented Building Blocks in that it delivers its findings to someone other than the people being judged. While Auditors should be polite and treat everyone professionally, the people they judge are not their customers. The epic failure of Arthur Anderson is an example of misunderstanding this. [33] And the financial crises in 2008 was, in part, caused by rating agencies who are paid by the banks whose securities they judge.

Audit sells its services to people other than those whom they inspect, such as the enterprise's Board or external entities.

To illustrate the difference, a testing service is not Audit. (It's a Service Provider.) Engineers can voluntarily use it (or not), and test results are reported back to the Engineers.

Similarly, an "assessment" is distinct from Audit. An assessment is voluntary (requested by a manager, perhaps to prepare him/her for real audits); and the results are reported back to those who were inspected.

Chapter 16: Audit 137

By contrast, an audit is imposed; and the results are reported to someone other than those being judged.

Audit's job is to catch people who are not in compliance with rules (such as laws, regulations, policies, standards, codes of conduct, and financial reporting), or to stop people from making mistakes by vetoing their decisions.

Need for Audit

Audit is often necessary, but it should be considered the *control mechanism of last resort.* It's far more expensive, and less effective, than systemic controls.

When the rules of the game induce people to comply — i.e., when compliance is in everyone's parochial self-interests — then control is systemic and Audit is unnecessary. For example, you don't have to force people to optimize their performance appraisals, or inspect entrepreneurs to ensure they're producing profits.

But when systemic controls aren't feasible, manual inspections and judgments are required.

One situation that illustrates the need for Audit is when customers can't perceive differences in product quality. For example, if everyone had a device that could identify any contaminants in food, there would never be another case of food poisoning. But because consumers cannot know if an item in a grocery store is contaminated, the government (in the US, the Food and Drug Administration) inspects food producers — an Audit function.

Of course, they can't examine every piece of food; so they

periodically inspect processes and test samples — an expensive and imperfect control. But it's better than nothing.

Another situation where Audit is needed is when employees have the ability to gain personally by sacrificing the well-being of the organization (e.g., where personal conflicts of interests might induce fraud). Where this occurs, if systemic checks-and-balances aren't practical, a control such as Audit is needed.

Scope of Audit

The mission of Audit is strictly to uncover problems.

It must not recommend corrective actions to those problems. Auditors' expertise in finding problems doesn't qualify them to design solutions (the expertise of other Building Blocks).

Furthermore, recommending a solution would be exercising undue influence, a conflict of interests. Imagine the Internal Revenue Service recommending that you solve your compliance problem by buying a particular brand of financial software!

Also, if Audit recommends solutions, it would no longer be "arm's length." It might judge more harshly a solution that complies but isn't what it recommended, or overlook problems just because people followed its recommendations.

Additionally, Auditors must not disempower managers by setting objectives or giving orders. To have legitimacy, the order to comply with rules and policies must come through one's chain of command, as should the directive to cooperate with Auditors.

Bottom of the T

Auditors are experts in the policies, regulations, or rules that they are judging. They're also experts in the audit process, i.e., how to collect and analyze the data they need to make their judgments.

Figure 15: Competencies of Audit

Bottom of the T: The topic being judged
Key methods: Audit process
Technical skill: High
Project management: Medium
Service management (operations): Low
Business of organization's customers: Low
Interpersonal skills: Medium
People supervision: Low

Audit is biased in favor of compliance, regardless of the needs of the business. Its biases are summarized in Figure 16.

Figure 16: Biases of Audit

- **Invention versus operational stability:** Operational stability
- **Purpose-specific solutions versus components:** Agnostic
- **Enterprisewide decisions versus specialized expertise:** Agnostic
- **Technical specialization versus unbiased diagnosis of clients' needs:** Specialization in the topic of their audits
- **Service oriented versus audit:** Audit

Audit is not only a distinct Building Block. To avoid conflicts of interests, it must be kept entirely separate from all the other service-oriented Building Blocks. If you're in a service business, you can't build an open, collaborative relationship with customers while also judging them.

SYNOPSIS

» Audit inspects and judges others, and reports the results to someone other than those being inspected.

» The mission of Audit is strictly to uncover problems, not to recommend solutions.

» Audit must be kept arms' length from all the other service-oriented Building Blocks.

~ PART 4 ~
Applying the Principles, Seeing the Problems

We live in a rainbow of chaos.

Paul Cezanne

Okay, now we're ready to look at any organization chart and anticipate where structure is getting in the way of people's performance.

The next Chapter introduces the method — one which puts the Principles and Building Blocks to work to diagnose an organization chart.

Then, Part 5 gives you a chance to practice the method by analyzing a series of case studies of actual structures which were designed to fail.

Parts 6 and 7 use this same method to help you design a new organization chart — one that's based on the Principles and Building Blocks, and tailored to the needs of your organization.

Chapter 17:
Diagnosing an Organization Chart:
The Rainbow Analysis

The first step in analyzing an organization chart is applying the Building Blocks to it.

The "Rainbow Analysis" does that, and then guides you through four questions to diagnose all the problems designed into your current organization chart. It reveals the strengths and weaknesses of your current structure — who's set up to fight with whom, and who's set up to fail.

The Rainbow Analysis can also be used to examine a proposed organization chart before it's implemented. If you're about to announce a new structure, this will greatly improve your odds of success; it may even help you avoid costly mistakes.

One caveat before we start: Identifying structural problems is not meant to accuse your organization of failing, or to make people defensive. Good people can overcome almost any structural dysfunction if they work hard enough. So the Rainbow Analysis only predicts potential problems, not actual failures. It highlights areas where structure is making it harder for people to succeed.

The Rainbow Analysis Workshop

The Rainbow Analysis is the basis of a highly interactive workshop in which a leadership team diagnoses its current organization chart. In this workshop, they can come to a consensus on whether, and how much, change is needed. And the experience helps them understand more deeply the science of structure.

Chapter 17: Diagnosing an Organization Chart: The Rainbow Analysis 143

I've facilitated the Rainbow Analysis workshop for dozens of leadership teams, in corporate, government, and not-for-profit enterprises. Invariably, it's an eye opener. Participants see their organizations and their jobs in new ways. And even if they don't decide to restructure, they understand where problems are coming from and why people are struggling.

This Chapter describes how that workshop works, in case you'd like to try it with your leadership team.

Of course, you can do the same analysis on your own. The steps are exactly the same.

(Rainbow Analysis Notes and Guidelines are available in the Supplement to this book; contact <ndma@ndma.com>.)

Data Collection

The process begins with an organization chart, such as your current structure or a proposed new structure. If you're doing the Rainbow Analysis in a workshop, print the organization chart in poster size and put it up on the wall.

Typically, it's sufficient to examine just two tiers of an organization chart. In large organizations, additional management layers may be needed. (Okay, it may take a few posters.)

Next, we'll need a set of eight or more colored marker-pens.

Considering one Building Block at a time, revisit the definition, and interpret it in the context of the mission of your organization.

Then, each leader takes a turn putting a colored stripe under his/her box on the chart if it performs that function. Put a red stripe under any box that's doing Sales work; a blue stripe under Applications Engineers; purple under Base Engineers; and so on.

In other words, color-code the organization chart, identifying which Building Blocks are within each group. (We often augment the colors by noting the Building Block and its sub-domain.)

If you're analyzing your current organization, be honest and color-code it based on what people actually do, regardless of what the organization chart says they're supposed to do.

In most of the organizations in which I've facilitated this process, the result is a very colorful chart. That's why I call it the "Rainbow Analysis."

This data-collection step helps build a deeper understanding of the definitions of the Building Blocks, essential to learning and applying the science of structure to your organization.

It also provides the data for analysis. Once the color coding (or labeling) is done, four questions will tell you where the problems are (summarized in Figure 17). Ask yourself if any of these are applicable to your organization:

Question 1: Gaps

A gap is any color (line of business) that's missing, or is done part-time by people whose primary focus is another function (a color that's spread around the chart in combination with other colors). It's nobody's primary job.

Of course, a color may be there but some sub-specialties within it may be missing. To find these gaps, look more closely at the specific sub-domains within each group marked with a given color.

Figure 17: Four Questions of the Rainbow Analysis

[Relevant Principles are cited in square-brackets.]

1. Gaps
   A. Unreliable processes [3]
   B. Reduced specialization [2]
   C. Overlaps [3]
2. Rainbows
   A. Reduced specialization [2]
   B. Conflicts of interests [5]
3. Scattered Campuses [6]
   A. Less professional exchange [6]
   B. Domain gaps (see Question 1)
   C. Domain overlaps [3]
   D. Less coordination [6]
   E. Not a whole business [1, 7]
4. Inappropriate Substructure [4]
   A. Reduced specialization [2]
   B. Domain overlaps [3]
   C. Inappropriate biases [4]
   D. Not a whole business [1, 7]

Where you have gaps, here's what you'd expect:

Unreliable delivery: With no one thinking about a line of business on a daily basis, it's an unreliable process. The organization misses opportunities (and it may not even know what it missed).

Reduced specialization: Work is done by people who don't specialize in that profession.

As a result of both these problems, the function is unreliable and ineffective.

Beyond that, gaps lead to overlaps when groups compete for control of missing domains (or just fill gaps for their own needs).

The problems with gaps are summarized in Figure 18.

Figure 18: Consequences of Gaps

[Relevant Principles are cited in square-brackets.]

A. Unreliable processes [3]: unreliable product/service delivery
B. Reduced specialization [2]: lower productivity, slower delivery (time to market), lower quality, greater risk, less innovation, more stress, and lower motivation
C. Overlaps [3]: reduced specialization, redundant efforts, less innovation, confusion, product dis-integration, territorial friction, lack of entrepreneurship

Question 2: Rainbows

"Rainbows" are easy to spot. They're groups marked with more than one color, delivering multiple Building Blocks.

Rainbows create two major problems:

Reduced specialization: In rainbow groups, people are expected to be experts at too many different things. Thus, they're mediocre at many of their assignments, and may neglect other duties altogether.

Conflicts of interests: A more serious problem results from the combination of incompatible Building Blocks.

For example, staff may be expected to keep operations stable (Service Providers), and also to innovate (Engineers).

Or they may be expected to specialize in a subset of the organization's products or services (Engineers or Service Providers); and at the same time, they may represent the entire organization to clients (Sales) where they're supposed to be completely unbiased.

Of course, the combination of Audit with any other Building Block is a serious conflict of interests. You cannot both judge people, and build a service-oriented partnership with them.

The problems with rainbows are summarized in Figure 19.

Figure 19: Consequences of Rainbows

[Relevant Principles are cited in square-brackets.]

A. Reduced specialization [2]: lower productivity, slower delivery, lower quality, greater risk, less innovation, more stress, lower motivation
B. Conflicts of interests [5]: domain gaps, unpredictable and uncontrollable balance of conflicting objectives, stress

Question 3: Scattered Campuses

When a color appears many different places on an organization chart, that's a "scattered campus." In the worst cases, all the different sub-specialties within a Building Block only come together at the level of the organization's top executive.

A scattered campus reduces professional exchange of experiences, discoveries, techniques, and work products. This leads to higher costs, slower delivery times, and product dis-integration.

It reduces coordination of the profession, such as weaker mentoring and management controls, less load balancing, limited career paths, confusion, and missed enterprise synergies.

And if a single line of business is divided, no one may feel accountable for managing that business. Worse, if one group's job is to oversee or decide how another group does its work, then neither is a whole business and they're both disempowered.

Over time, a scattered campus often leads to gaps and overlaps, since no one manager (short of the top executive) is in a position to adjudicate domains.

The problems of a scattered campus are summarized in Figure 20.

Figure 20: Consequences of Scattered Campuses
[Relevant Principles are cited in square-brackets.]

A. Less professional exchange [6]: higher costs, slower delivery times, product dis-integration

B. Less coordination [6]: weaker mentoring and management controls, less load balancing, limited career paths, confusion, missed enterprise synergies

C. Domain overlaps [3]: reduced specialization (lower productivity, slower delivery, lower quality, greater risk, less innovation, more stress, lower motivation), redundant efforts, confusion, territorial friction, product dis-integration, missing entrepreneurships, disempowerment

D. Domain gaps [3]: unreliable delivery, reduced specialization (lower productivity, slower delivery, lower quality, greater risk, less innovation, more stress, lower motivation)

E. Disempowerment [1, 7]: lack of customer focus and entrepreneurship, less business planning, disempowerment (unconstrained controls, wasted talents, focus on tasks instead of results, less innovation, inefficiencies and ineffectiveness, unfair scapegoating, unclear accountability for results)

Chapter 17: Diagnosing an Organization Chart: The Rainbow Analysis

Question 4: Inappropriate Substructure

To analyze the final question, look at both the colors and the words in the boxes. The color (Building Block) tells you what a group's specialty is supposed to be. Now scan across the organization chart to see how that domain is subdivided. If the basis for substructure is anything other than what the Building Block is supposed to be good at, that's an inappropriate substructure.

For example, you might see Engineers divided into groups based on the clients' organization chart, or business processes, rather than the types of products they produce. Or you might see Service Providers divided up based on the technologies they use rather than the services they provide.

Figure 21 summarizes the appropriate bases for substructure.

Figure 21: Bases for Substructure

Services Providers:	Service
Engineers, Application:	Purpose *(in IT, data object)*
Engineers, Base:	Engineering discipline
Coordinators:	What they coordinate (broadly classed into business, technical)
Sales, Account:	Client territories (business units)
Sales, Retail:	Geography, venue
Sales, Function:	Client profession or business process
Marketing communications:	Communications channel, service
Marketing research:	Service *(e.g., design, data collection, analysis)*

The wrong basis for substructure reduces specialization, which results in lower productivity, slower delivery, lower quality, more risk, less innovation, more stress, and poor morale.

It generally creates overlapping domains, which creates confusion, causes redundant efforts, undermines product integration, creates territorial friction, and diminishes entrepreneurship.

It may also induce inappropriate biases. For example, a Sales function divided by product line would give clients product-driven (rather than purely business-driven) recommendations. Essentially, the structure leads people to optimize the wrong objectives.

Worst of all, if groups are divided by anything other than a line of business — such as by tasks, or where one group's job is to oversee another — there are serious consequences of disempowerment.

Figure 22 summarize the risks of the wrong bases for substructure.

Figure 22: Consequences of Inappropriate Substructure

[Relevant Principles are cited in square-brackets.]

A. Reduced specialization [2]: lower productivity, slower delivery, lower quality, greater risk, less innovation, more stress, lower motivation

B. Domain overlaps [3]: reduced specialization, redundant efforts, less innovation, confusion, product dis-integration, territorial friction, lack of entrepreneurship

C. Inappropriate biases [4]: poor advice, optimizing the wrong objectives

D. Disempowerment [1, 7]: lack of customer focus and entrepreneurship, less business planning, disempowerment (unconstrained controls, wasted talents, focus on tasks instead of results, less innovation, inefficiencies and ineffectiveness, unfair scapegoating, unclear accountability for results)

Interpreting the Rainbow Analysis

When I facilitate leadership team workshops, as we go through the four questions, we discuss each potential dysfunction, and take notes of the relevant problems (those which really seem to be affecting the organization). We list specific examples, such as exactly which sub-specialties are missing, or where domains overlap.

Then, considering that list of relevant problems with the existing structure, I raise the next question: *Is change needed? And if so, do you need just a few small "tweaks" or a "clean sheet of paper" approach?*

The choices at that point are:

A. Do nothing
B. Tweaks: a few small changes to the existing structure
C. Clean sheet of paper: a completely new structure

Be forewarned: As discussed in Chapter 29, many small tweaks are more difficult than a Clean Sheet approach; and they're typically less effective, lacking the whole-system perspective. Unless the current structure is very close to right, a clean sheet of paper is generally the right way to go.

If you're doing this on your own (rather than in a workshop with your leadership team), I encourage you to follow the same steps. Take notes as you analyze each question, and then assess how serious your organization's problems are.

SYNOPSIS

» The "Rainbow Analysis" identifies the lines of business in each group.

» A "gap" is any line of business that's missing, or is done part-time by people whose primary focus is another function. The missing function is likely to be unreliable and ineffective.

» "Rainbows" are groups fulfilling multiple Building Blocks. They reduce specialization, and may introduce conflicts of interests.

» A "scattered campus" is a line of business reporting to many different groups. It reduces professional exchange and management coordination, and may lead to gaps and overlaps (or in some cases, disempowerment).

» "Inappropriate substructure" is where a function is divided into groups by something other than its specialty. It reduces performance, introduces inappropriate biases, and may indicate disempowerment.

» After listing the problems, you'll be positioned to decide if change is needed, and how much.

~ PART 5 ~
Structures Designed to Fail

Study the past if you would divine the future.

Confucius

Before you start drawing boxes for your own organization, it may be useful to practice a bit. The Chapters in this Part are a series of case studies that describe approaches to structure that others have tried — some of which you might find familiar — and we use the Rainbow Analysis to analyze their faults.

These case studies give you practice looking at organization charts and anticipating their problems. They also warn you about approaches that may be popular but have some serious flaws, so that you won't risk repeating mistakes others have made. (Note that we've already diagnosed a number of flawed structures. For a list, see the table of Case Studies on page *ix*.)

Again, many of the case studies use IT as the example, to provide a consistent environment in which you can see the Principles and Building Blocks at work. And again, we trust you can apply the lessons to your own industry or function.

Hint: Seeing the Rainbow Analysis in action will help you apply the science of structure to your own organization. But you won't miss any new content if you skip some Chapters in this Part which don't interest you.

Chapter 18:
Strategy as a Basis for Structure

This case study explores a long-standing misconception of the relationship of structure to strategy.

Situation

In 1962, Alfred D. Chandler put forth the well-known proposition that "structure follows strategy." [34] He theorized that an organization's structure could be fine-tuned to accomplish its well-defined strategy.

Some still believe this. For example, in a press release dated June 17, 2015, Satya Nadella, CEO of Microsoft, said: *"I'm certain that matching our structure to our strategy will best position us to build products and services our customers love and ultimately drive new growth."*

On the surface, it may seem logical to organize around strategies. But there are serious flaws in this reasoning.

Analysis

Here's what you'd see in the Rainbow Analysis:

Gaps: "Strategy" is rarely singular. Enterprises pursue multiple strategies, each of which has many facets. Not every strategy can be represented in the structure. There inevitably are gaps.

And strategies are ever changing. When a new strategy, or variant of a strategy, occurs, there may be no group dedicated to it.

With such gaps, some strategies are either ignored or poorly executed.

Rainbows: This approach creates a "silo" organization, where each strategy-centric group includes all the professional specialties it needs to achieve its objectives. The reduction in specialization hampers every strategy.

Scattered campus: Of course, many strategies draw on the same specialties. If an organization divides its talent into groups by strategy, specialties will be scattered around the organization. The overall enterprise strategy is likely to become disjointed; and economies of scale and synergies are lost.

Inappropriate substructure: Staff are encouraged (by the nature of the structure) to specialize in an enterprise strategy, rather than in the competencies they're supposed to contribute to that strategy. This is an inappropriate basis for substructure. Again, reduced specialization hampers performance.

Common sense: There are also some practicalities that argue against this approach. The pace of business has changed. In Chandler's day, business strategies may have been relatively stable and long-term. But no more! Strategies can, and do, shift quickly and continually. To remain competitive, organizations must learn to be flexible and responsive — to "turn on a dime."

If structure were to follow strategy, organizations would have to restructure on a frequent basis. This, of course, would be prohibitively disruptive, expensive, and disconcerting to staff. And it would delay response to new strategies.

But if they don't continually restructure, organizations tuned to do an excellent job of today's strategies are destined to perform poorly at tomorrow's strategies.

Perhaps worse, an organization structured around today's strategies will fail to *discover* tomorrow's strategies. [35]

Alternative

A well-designed organization chart, combined with effective teamwork, can handle any strategy. Here's how it should work:

A strategy translates into specific initiatives.

For each initiative, clear domains identify the one and only group that "sells" that product or service. This becomes the "prime contractor" for that strategic initiative.

Then, teamwork processes take over (Part 8). That prime contractor may get help from any of the other groups anywhere in the organization, and they in turn from still others.

This way, each strategic initiative draws from the existing structure all the needed competencies. And the organization can quickly recombine its talents on new teams to address an unlimited number of strategies.

This way, a healthy structure responds effectively to a continually changing mix of strategies; and yet it provides the stability that allows people to focus on their specialties, which cultivates excellence.

Beyond that, in an entrepreneurial organization, everybody continually develops their own strategies as entrepreneurial businesses within the business.

In summary, healthy organizations flexibly respond to diverse client strategies, while continually generating their own strategies — reversing the old adage that structure follows strategy.

SYNOPSIS

» When strategy is used as the basis for structure, an organization is less effective at current strategies, and it may fail at future strategies.

» A well-designed structure is adaptive to constantly changing strategies, and continually generates its own strategies.

Chapter 19:
Pick a Core Competency, Outsource the Rest

Here's another misguided translation of strategy into structure.

Situation

Michael Treacy and Fred Wiersema posited that there are three "competencies" that organizations require: operational excellence, product/technology leadership, and customer intimacy. [36] They explained why companies (or business units) must choose just one on which to base their market reputation and strategies.

Although Treacy and Weirsema caution against abandoning the other two, some theorists recommend that organizations optimize their structure to focus on just one.

For example, Jay Galbraith contrasts a product-centric firm and a customer-centric firm, [37] as if a company cannot be both (by having both a product-centric Engineering organization and a customer-centric Sales and Marketing organization, collaborating via teamwork).

Why Not Outsource the Other Competencies?

In this same vein, innumerable vendors and consultants suggest that you outsource everything not considered a core competency, to the point where this axiom has become common wisdom. [38]

Outsourcing entire functions isn't effective for many reasons.

First, note that outsourcing less critical competencies doesn't magically strengthen the remaining competency. It doesn't leave any more people focused on the chosen competency.

And managing vendors is certainly not easier than managing staff. Vendor contracts are often difficult to exercise, distracting executives with complex legal negotiations.

Furthermore, even in less critical (but still needed) competencies, outsourcing can damage performance. It can reduce flexibility, since any unanticipated requirements generally must be negotiated, contracted, and come at a high price.

Another reality is that vendors can never be as well aligned with business strategies as insiders. Despite the talk of partnership, vendors are morally obliged to optimize the best interests of *their* shareholders, not yours.

Even when their vested interests are aligned with yours, vendors don't contribute to your strategies in the same way. They do as they're told (in their contracts). Despite talk of "partnership," they never contribute to strategic thinking at the same level as leaders within an enterprise.

And in many cases, outsourcing doesn't save money. In fact, just the opposite, paying vendors a profit to do what you otherwise could do can turn out to be expensive.

So ultimately, outsourcing an entire function weakens it. Before we discuss the value of outsourcing and a better way to utilize vendors, consider the Rainbow Analysis.

Analysis

Here's what you'd see in the Rainbow Analysis if you outsource an entire function (competency):

Gaps: An organization designed around a single competency sacrifices its effectiveness at the other competencies. This is risky for two reasons:

First, most strategies require a mix of competencies. Weakness in any competency can jeopardize the strategy.

For example, a strategy focused on customer intimacy requires operational excellence in points of customer interface, such as customer service and web applications.

Innovation strategies such as product development require input from sales and marketing (customer intimacy) and manufacturing (operational efficiency).

And operational efficiency strategies are built on excellence in engineering (innovation).

Second, business conditions are volatile, so strategies must be dynamic. For example, a dip in the economy may require a shift in emphasis from innovation to operational efficiency, to be reversed when the economy recovers. An organization designed to pursue only one competency is ill-equipped to respond when business imperatives require other strengths.

For both these reasons, no organization can afford to focus exclusively on just one competency, and settle for mediocrity in the others.

Alternative

For each Building Block, one competency is dominant. Sales and Marketing is dedicated to customer intimacy; Engineering to innovation; and Service Providers to operational efficiency.

When diverse groups are combined into an organization, the organization as a whole can include all the Building Blocks — and all the competencies. So there's no reason why an organization can't excel at all those competencies. In addition to enabling any strategies, this ensures that there's a voice for each perspective in the strategy-formulation process.

When To Use Outsourcing

Although outsourcing is not a good substitute for a strong internal function, vendors can add a lot of value. In some cases, they may be cheaper, better, more flexible, and more responsive.

Specifically, external vendors are more cost-effective when: [39]

- Multiple corporations can share a vendor's assets. This only has value when economies of scale cross corporate boundaries. An example is telecommunication networks, where many companies buy services from vendors so as to share long-distance lines.

- Due to its size, a vendor can afford a higher degree of specialization. This is particularly valuable in high-tech professions where only the largest organizations can afford to hire a qualified individual. Or applying this criterion to a service, volumes may be so low that an organization cannot afford to produce it internally. In these cases, outsourcing provides capabilities the organization cannot afford on its own.

- A company requires more capital than is available to it; so it's willing to pay a premium in operating expenses to <u>use other companies' capital</u>.

- Business volumes vary widely, or grow more quickly than an organization can hire staff and acquire resources. So it's worth paying more to <u>make costs variable rather than fixed</u>.

How to Integrate Vendors into Your Organization

So what's the appropriate way to take advantage of vendors without giving up any internal competencies?

Vendors should never be used as an *alternative to* internal staff. Instead, vendors should be *hired by* internal staff to extend their productive capabilities or improve their value proposition.

I call this approach to managing vendors "**extended staffing**." Essentially, **vendors are part of the staff of the appropriate internal function**, not an independent service provider that works directly with other functions. Clients get the product or service from internal staff, who may use vendors as part of their delivery capability.

This should occur naturally in an entrepreneurial organization. Entrepreneurs don't focus on growing empires (costs and headcount). Rather, they strive to be the supplier-of-choice to their customers. If it's more economic to "buy" than "make," they're the first to propose it. That way, they always offer the best deal.

This is not "brokerage," as some have proposed. Brokerage

implies just connecting a buyer and a seller, with no accountability for the deliverables. Instead, internal staff are "value-added resellers" of vendor services. People who know the profession (and who work for your shareholders/taxpayers/donors) are the ones who make commitments to clients, manage the vendors, and retain accountability for results.

The services that internal entrepreneurs offer their customers may not precisely match those that vendors sell to them. Staff may add value to vendor-services to better meet the needs of internal customers (such as integration, security, business continuity, compliance, and ongoing support). They may assemble multiple vendor-services into a more complete service. Or they may just repackage a service, and perhaps charge for it (distribute costs) in a different way.

In any case, extended staffing promotes tight vendor integration. Internal staff oversee vendors to ensure performance and compliance with internal policies, standards, and plans; and they incorporate them in internal service-delivery processes.

Of course, internal staff incur costs in managing vendors (at least their own time). But since they're the specialists in their professions, it's more effective and efficient for them to manage vendors than for consumers to do so.

Another advantage of extended staffing is its dynamic nature. When industry economics shift in either direction, staff can utilize vendors more or less. Thus, the enterprise gets the best balance between "make" and "buy" at any point in time.

Extended staffing gives enterprises the benefits of outsourcing, without forsaking internal competencies.

So with different Building Blocks fulfilling each of the competencies, and with extended staffing augmenting staff's capabilities, there's no need to choose just one core competency and outsource the rest. Enterprise strategy may focus on just one; but your structure should encompass all the competencies.

SYNOPSIS

» Business strategy may focus on just one of the three competencies that organizations require: operational excellence, product/technology leadership, and customer intimacy.

» But structure should contain all the competencies. Completely outsourcing any of the three can damage the others.

» Different parts of an organization (Building Blocks) specialize in each. Then, teamwork permits the organization to excel at all.

» Vendors should be used to augment the capabilities of internal staff ("extended staffing"), not to replace them.

Chapter 20:
Managers as Client Liaisons

This case study describes a common mistake made by internal service providers.

Situation

In one company, client executives complained that Corporate IT didn't understand their strategies, their people, or their needs. When they demanded a single point of contact, the CIO assigned a set of business units to each of his senior leaders. Thus, each leader had two roles:

Role #1: Manage a technology-engineering or service group, ostensibly available to all clients.

Role #2: Serve as liaison to a set of client business units, ostensibly representing the entire IT department.

Phil headed up a group of IT applications developers for financial systems such as the general ledger, accounts payable, and tax. So he was assigned to the Finance department — not for just his own team's projects, but for all projects and services delivered to clients in Finance. He was considered the "relationship manager" on behalf of all IT to the CFO.

His peers were given similar dual roles for other applications domains and clients. Sam, for example, managed customer applications and was the liaison to Sales and Marketing clients.

And Bev managed human resource applications (employee data) and was the liaison to the HR department.

In this IT department were the following groups (among others):

- Phil, financial applications   + Finance clients
- Sam, customer applications   + Sales and Marketing clients
- Bev, employee applications   + Human Resources clients

Analysis

Here's what you'd see in the Rainbow Analysis:

Gaps: Clients rightly want their IT liaisons to be readily available. The Sales function should spend most of its time with clients.

In terms of time commitment alone, this is a full-time job. But Phil had a big group to manage. He couldn't spend sufficient time with clients attending their meetings; getting to know the people and their businesses; working with them to discover new opportunities; brokering clear agreements; delivering account reviews; resolving relationship issues; etc.

Beyond that, the Sales role requires specialized skills and methods, especially critical if this function is to do more than facilitate relationships and take orders. One example is a method to discover really strategic opportunities.

Phil knew that to do this function well would require study and experience. This was not a matter of capability. He felt confident that, if given proper time and training, he could learn to perform the client-liaison job well. But while he did so, who would manage his applications development group? Phil just didn't have time to do both.

In truth, even if he had the time, Phil didn't have the right background. He was an IT expert, not a finance professional. He understood their accounting processes; but he didn't really understand their business challenges and strategies.

Most of the time, Phil remained focused on his original Engineering job for a number of reasons. Project pressures demanded it. Technology was his first love and the focus of his career. And the people in his group depended on his guidance, coaching, and leadership.

Thus, the Sales function remained a gap. These senior managers did provide a point of contact which improved relations. But they didn't do much to add value to clients' strategic challenges.

Rainbows: To the extent that he did spend time with clients, most of Phil's interactions were with the Accounting department, the primary users of his financial applications.

However, Phil couldn't help but notice that his traditional clients were some of the least influential of the Finance department's senior management team. He saw that Jane, the manager of a small Financial Planning and Analysis group reporting directly to the CFO, seemed to get a lot more attention.

So Phil called Jane. He told her he was looking for high-payoff, strategic IT projects, and wondered if he could serve her in any way. Jane invited Phil to drop by.

When he did, Phil found Jane in the midst of analyzing staffing cost trends. What she really needed was access to employee data, market data, and data-analytics tools.

But what did Phil recommend? Can we honestly expect this manager of financial applications developers to prescribe anything other than extensions to his general-ledger application!? Despite his best intentions, technological bias crept in; and the project that Phil suggested was only of mild interest to Jane.

This structure put Phil in a conflict-of-interests situation. As an expert in a set of technologies (financial applications), he was paid to be biased — the natural bias of a specialist. At the same time, in the client-liaison role, Phil was expected to be unbiased and represent the entire IT product line.

No matter how well-educated and well-intentioned Phil may have been, organizational forces were working against him. It wasn't that Phil made a conscious decision to be parochial. Phil honestly believed that the financial solutions to which he was dedicated were the greatest thing since sliced bread, and he just didn't see opportunities outside his own specialty.

This rainbow of Sales and Engineering led to product-driven, rather than business-strategy-driven, recommendations.

This conflict of interests was not lost on Jane. She found it difficult to trust the objectivity of someone who also had responsibility for (and hence a vested interest in selling) one particular solution.

Inappropriate substructure: In this case, Jane persisted and got Phil to recognize her need for a solution other than financial systems. But then he ran into another problem. The other groups who supplied needed data and tools were busy serving their own clients.

For example, Jane needed employee data. But Bev's clients in HR

Chapter 20: Managers as Client Liaisons 169

set priorities for her staff. Jane's Finance project didn't make their list.

Of course, Phil still had to find a way to get the job done. When he couldn't get help from Bev, his group developed a small HR application (outside their domain of expertise).

Similarly, clients outside the Finance department had trouble getting help from Phil's group, since his priorities were set by the CFO. As a result, multiple overlapping financial applications cropped up throughout the company. This structure led to costly replication of efforts and dis-integration of data.

Ultimately, the structure devolved into client-dedicated Engineering groups (as discussed in Chapter 4). Productivity and quality suffered. And enterprisewide synergies were lost.

Alternative

This mixture of clients and technologies was an attempt to fill a gap in the structure: the Account Sales function.

But Sales is not a part-time job for an internal service provider's senior leaders, any more than engineering and manufacturing executives can serve as a corporation's sales force in their spare time. Sales should be a separate group, within internal service providers as it is within companies.

Some may say that their organization can't afford it. But the truth is, a small, dedicated team of Sales professionals will perform far better than the same number of person-hours spent on client-liaison work distributed among the managers of other functions.

And some may say that their clients won't accept a "sales"

function in an internal shared-services organization. Okay, so give it a different name. But in addition, define the services that Sales delivers (Chapter 15) to help clients understand its value to them.

The bottom line is: Sales is a profession in its own right, and it warrants dedicated specialists.

SYNOPSIS

» Making Sales a part-time job for an internal service provider's senior leaders leads to product-driven recommendations, reduced specialization, replication of efforts, and lost synergies.

» Sales is a profession in its own right, and should be a distinct group within the structure.

Chapter 21: Decentralization

The Rainbow Analysis reveals why decentralization, including the "federated model," drives costs up while producing lower quality.

Situation

Business unit leaders may prefer to have their own support staff rather than work with an enterprise shared-services department; i.e., they choose decentralization.

The allure typically comes down to these five things:

- **Customer focus:** Their own staff will respect them and treat them well; but shared-services providers may treat them as a nuisance rather than a customer, ignore their requests, or attempt to control rather than serve them.

- **Understanding their business:** Their own staff are "closer to the business" and understand their unique needs, whereas some shared-services providers believe that "one size fits all" and force on them solutions that don't fit their needs.

- **Control of priorities:** They can control the priorities of their own decentralized groups; but some shared-services providers make them wait in line, beg for attention, or work through committees and other bureaucratic hurdles to get what they need.

- **Control of costs:** If shared-services providers allocate their costs, business unit leaders can control their expenses more easily when the group reports to them.

- **Accountability:** There's also the mistaken belief among some corporate executives that decentralization is required in order to hold business-unit leaders fully accountable for their results — as if doing business with a vendor (internal or external) diminishes their authority over, and hence accountability for, their business performance.

Analysis

Of course, decentralization de facto substructures all the Building Blocks by client (like the case study in Chapter 4). The Rainbow Analysis reveals the problems with this:

Gaps: Managers of decentralized support groups are still not effective at the Account Sales function, being busy managing their groups. For example, they rarely study the methods needed to discover new strategic opportunities. Result: less strategic value.

And decentralized groups do little to improve collaboration with shared-services providers. In fact, they often see their role as defending their business unit *against* the corporate function. Result: strained relationships.

Scattered campus: Each specialty is scattered among the business units. Small groups that have to cover many domains can't specialize to the same degree as a consolidated function. Result: lower performance, less innovation, and higher costs.

Inappropriate substructure: Every specialty is substructured by client. So each builds parochial solutions, undermining enterprise synergies.

Enterprise synergies can be found in every internal support function. Shared information (IT), talent management (HR), space

Chapter 21: Decentralization

(Facilities), product standards and parts (Engineering), external customer touch-points (Marketing), and production facilities (Manufacturing) all can create shared or compatible solutions that encourage collaboration across business units.

Decentralization undermines that collaboration among business units. Here's one case example:

In the spirit of autonomy, each division in a global heavy-equipment company built its own manufacturing plant. This increased costs for a number of reasons:

Capacity sat idle in one plant while another was pressured beyond its capacity. Ultimately, lots of excess capacity ("safety stock") was built into the system.

One division had a plant in a country while another division did not. Instead of using that local plant, the other division shipped its products from its own plant in a neighboring country, and transportation costs rose.

One division pioneered new manufacturing techniques, but other divisions didn't learn about them until much later when the CEO intervened and commissioned a corporatewide task-force to rationalize production capacity.

And while there could have been significant economies if some plants were designed around long, stable manufacturing runs while others were designed for rapid re-tooling, none of the divisions alone could afford to specialize to that degree.

This same company provided another case example in their engineering function:

Each division also had its own design engineering function. With all the reinvention, the number of parts the corporation had to make or buy skyrocketed.

That same CEO-sponsored task-force found a dozen different electric motors with roughly the same specifications. Costs of development, manufacturing, inventories, and support all went up.

I personally experienced the effects of decentralization of IT as a customer of an insurance company in this <u>case</u> example:

The decentralization of its IT function resulted in multiple customer databases. Since customer-numbers varied, the enterprise was not able to spot customers who bought from multiple business units.

The Specialty Automotive business unit cancelled the policy covering my vintage sports car, saying they were no longer interested in that type of business. Annoyed, I moved <u>all</u> my insurance to another company — the policies for my other cars, my home, my personal liability umbrella, and my corporation.

Of course, they never knew this. Because their customer databases were fragmented among their business units, they knew me as a few discrete policies, not as one customer with diverse needs.

Okay, I'm not a big customer; but I'm sure there are others like me. This is just an example of how decentralization of a function can undermine synergies across an enterprise.

The bottom line is, decentralization inevitably results in higher costs, slower delivery times, lower quality, less innovation, reduced career opportunities for staff, and fragmented products and services which undermine enterprise synergies. [40] (The only thing worse than decentralization is a shared-services organization that's not customer focused; doesn't understand the unique needs of each business unit; and denies clients the ability to control how much they spend and what they buy.)

Federated Model

In companies where decentralization is prevalent, there may be a weak enterprise function for just the obvious common services. All other services are left to the decentralized support groups. This is sometimes termed a "federated" model.

Of course, a nice name doesn't solve any of the problems caused by decentralization. It's little more than a truce, such that shared-services and decentralized functions aren't competing for territory.

Adopting a federated model actually makes things worse. It institutionalizes decentralization. Ideally, shared-services providers *should* compete for business by offering business units a better deal whenever possible. This competition doesn't have to undermine local authority. Shared-services providers can work through their decentralized counterparts, as suppliers to them.

Dotted Lines from Decentralized Groups to a Corporate Shared-services Leader

Some companies attempt to ameliorate the problems caused by decentralization with a "dotted line" drawn from decentralized groups up to the leader of the enterprise-level internal service provider. For example, a CIO might have dotted-line authority over decentralized IT groups that report to business units.

From a cynical perspective, this might be an attempt by top executives (like a CEO) to make corporate executives (like a CIO) accountable for the behavior of people they cannot control, in the spirit of holding one person accountable for the entire function. Imagine a CEO saying to a CIO, "It's your job to manage the entire IT function (including decentralized staff)." This violates the Golden Rule: accountability without real authority.

In response to such an untenable situation, corporate leaders may try to claim a degree of supervisory authority over decentralized staff. They may grant that business units can decide *what* is to be done, but they try to decide *how* work is done (professional practices). Some even attempt to manage career paths and contribute to performance appraisals.

Of course, any attempts to exercise authority over people who don't report to you — such as giving them assignments or dictating how they do their jobs — results in political struggles. Their real managers naturally fight any incursion on their authorities, as per the Golden Rule. They see this use of the dotted line as reducing their ability to meet their obligations.

And with their close relationships with local business-unit executives, the real (decentralized) managers generally win political battles. It's easy to imagine them saying, "Look, boss, if you want me to achieve your objectives, you've got to get this corporate guy off my back!"

Meanwhile, decentralized staff are caught in the middle. Of course, it's not fair to subject any employee to potentially conflicting directions and priorities.

Pragmatically, a dotted line gives shared-services executives no real authority. Decentralized staff get their funding from their business units, answer to business-unit leaders, and are quite willing to defend their business units against corporate meddling.

Holding a shared-services executive accountable for things he/she cannot control just induces strained relationships, and sets up that executive to take the blame when business units misbehave. To be effective, a CEO must direct staff through legitimate lines of authority — through business-unit executives — and not expect shared-services leaders to do his/her "dirty work."

Other Weak Palliatives

Other attempts to overcome the costly and damaging forces of decentralization are also ineffective. For example, "governance" in the form of committees and working groups adds bureaucracy, but it doesn't change the fundamental drivers of dis-synergies.

Another weak palliative is "centers of excellence," where decentralized groups in one business unit are expected to support others. Essentially, they become shared-services providers, but reporting to business units rather than at the enterprise level.

Of course, the "home" business unit controls their priorities; and the resulting poor service to others causes other business units to ignore centers of excellence and replicate the function.

Roles of Shared Services Amidst Decentralization

If you have to tolerate a decentralized (federated) environment, enterprise shared-services organizations can still fulfill five roles:

1) **Full-service provider to clients in corporate headquarters.**

2) **Sole (monopoly) provider of a short list of products and services** where the synergies and economies of scale are widely accepted. Often, these are services based on large, expensive assets (such as a manufacturing plant or data center); or

they may be services that link everyone in the enterprise (such as a network).

The products and services on the short list should be determined through a consensus of business-unit leaders, not imposed by corporate executives, so as not to further strain relations.

3) **"Outsourcing" supplier** whenever decentralized groups wish to buy from it. In this case, the decision as to what a shared-services provider will do for a business unit is completely at the discretion of the business unit, and shared-services staff must earn the business through excellent performance, value, and relationships.

By selling through (not around) decentralized counterparts, shared-services staff are not disempowering autonomous business units in any way. Yet as they earn market share, they can deliver some economies of scale and synergies.

4) Sole provider of **coordination services** where decisions must be made that affect the entire enterprise. Coordinators have no formal power; their job is to facilitate a consensus among stakeholders on standard, policies, and plans.

Shared-services staff can also facilitate collaboration among business units with common needs. Examples include shared vendor contracts (a purchasing service), and consortia where business units share a product or service.

Note that both these coordination and facilitation services are just that — services. As such, business units may or may not choose to participate. It's up to the CEO and business-unit executives to motivate stakeholders to collaborate, for example, by demanding enterprisewide standards and policies, or common business processes. Only then will decentralized staff choose to utilize these coordination and facilitation services.

5) **Spokesperson for the profession,** promoting the best interests of all staff throughout the enterprise. The key here is to never speak for others or make commitments for others, or attempt to manage staff who report elsewhere. Rather, a shared-services leader can further the profession by encouraging collaboration (professional interest groups) and representing the profession's interests in enterprise policy discussions.

By accepting accountability only for these five roles, a shared-services organization can contribute to enterprisewide objectives without threatening business unit autonomy, antagonizing potential customers, or becoming a scapegoat that takes the blame for problems engendered by decentralization.

One more benefit: By always treating decentralized counterparts in a respectful, customer-focused manner, relationships improve and shared-services leaders will find themselves with more, not less, actual influence. Sometime the "soft" approach is actually the strong approach.

But remember, despite these services, most of the costs of decentralization remain.

I'd rather have a customer than a hostage!

Preston T. Simons

Alternative: Shared Services

Ultimately, all these patches are only marginally effective. The right answer is consolidation of shared services.

"Shared services" refers to an internal service provider that serves multiple clients. Instead of each business unit owning its own little group, clients buy products and services from a central supplier.

There are many **benefits to shared services**.

Costs are reduced when redundancies are eliminated. A shared-services organization can eliminate parallel training, product R&D, policy formulation, and support functions.

Shared services does not mean that "one size fits all." A centralized (shared-services) team of specialists can tailor solutions to unique customers' needs at a lower cost and higher quality than small groups of decentralized generalists.

But with that said, another benefit is that a shared-services provider can spot opportunities for "consortia" where multiple clients share a single solution. Even if different business units require unique solutions, a shared-services provider can at least reuse some components, and can certainly reuse their knowledge and competencies.

Furthermore, consolidation offers economies of scale in both staff and infrastructure by better balancing workloads. One business unit's peak load may occur at a time when other business units are slow. Total peak demand is generally less than the sum of each business unit's peak. In the terminology of inventory management, centralization reduces the need for "safety stocks."

Centralization also consolidates buying power. Vendor licenses may be shared at a lower cost. And a bigger organization has more bargaining power and can drive a better deal.

Beyond just cost savings, sheer size improves performance. A larger organization can support a broader, more diverse product line. It can support its products better globally, 24 by 7. And it can afford higher-caliber management.

Also, substructuring staff in the appropriate way (rather than by client) increases specialization, with performance improvements such as lower cost, higher quality, and more innovation.

Perhaps the greatest benefits of consolidation come from the potential enterprisewide business synergies. As business units find themselves using the same services, they may collaborate more (and better) across boundaries. The implications for enterprise performance are as profound as the reasons why these business units are under the same corporate umbrella in the first place.

The bottom line is, a well-designed and well-managed shared-services organization not only performs better, but can improve performance throughout the enterprise.

Prerequisites for Shared Services

Before business units give up their decentralized groups, a shared-services organization has to address the reasons clients like decentralization. Consider the five problems that drive decentralization, all of which can be more effectively addressed in other ways:

Customer focus is a matter of culture — the habits and practices within the organization. It can be addressed by teaching staff the specific behaviors that embody the spirit of customer focus, including respect for customers' purchase decisions and a willingness to tailor solutions to clients' unique needs. [41] But first, the structure has to clarify whom their customers are.

Understanding of clients' businesses is the job of the Account Sales function. The answer is to dedicate an Account Sales professional, not Engineers, to each client business unit.

Control over priorities: Decentralization does give clients an understanding of the limits to their resources, and control over their priorities. But there's a better way to address this resource-governance challenge. Clients can be given control of spending power rather than dedicated staff, as described in Chapter 30. (Internal chargebacks are not necessary.) [42]

Control of costs: The costs of shared services should never be allocated to business units on any basis other than utilization. Holding business-unit executives accountable for costs they can't control is futile and unfair.

Accountability: Business unit leaders must be held accountable for their use of shared services, just as they're accountable for their purchases from vendors. To support this, shared-services organizations must be able to portray their costs by business unit, based on consumption. Until they can do so, shared-services costs must not be part of business units' cost structures.

If it meets these requirements, a shared-services organization doesn't threaten business unit autonomy any more than does buying from external vendors (which are shared across corporations). There should be no need for decentralization.

Chapter 47 describes the process of consolidating decentralized groups into a shared-services organization.

Chapter 21: Decentralization

SYNOPSIS

» Decentralization reduces specialization, and hence reduces performance and increases costs.

» Decentralization undermines enterprise synergies.

» "Federated" models and dotted lines from decentralized groups to a shared-services executive add confusion without solving the underlying problems of decentralization.

» There are five functions that shared services can fulfill to partially ameliorate the costs of decentralization; but these certainly don't solve the big problems.

» A centralized (shared-services) internal service provider can tailor solutions to unique customers' needs at a lower cost and higher quality than small groups of decentralized generalists.

» Before centralization can work, a shared-services provider must earn the position of supplier of choice by satisfying five requirements: It must be customer focused. It must have a Sales function that understands clients' unique needs. It must implement resource-governance processes that empower clients to choose what they "buy" from it (its priorities). It must never allocate costs on any basis other than utilization. And ultimately, it must be able to portray its costs by business unit based on consumption.

Chapter 22:
Dotted Lines and Matrix Structures

Sometimes clients who would prefer decentralization, but can't have it, demand a "dotted line" over shared-services staff.

The term "dotted line" means that one manager has some managerial authorities over, and perhaps accountabilities for, people who report administratively (solid line) to another manager.

In extreme cases, there may be two solid lines where two different managers supervise the same person, i.e., dual reporting.

Figure 23: Two Types of Dotted Lines

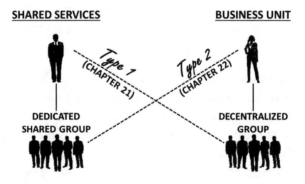

There are two types of dotted lines (portrayed in Figure 23):

1. Chapter 21 discussed a dotted line from decentralized groups up to a shared-services leader, in an inappropriate attempt to ameliorate some of the problems of decentralization.

2. This Chapter examines the opposite direction: dotted lines

from a group in shared services up to a business unit executive. The reasons commonly given are to reinforce customer focus and enhance relationships.

The group may be just the Account Sales function (which is appropriately dedicated to a set of clients); or it may be an entire client-dedicated group (case study in Chapter 4).

Problems Created by Dual Reporting

Whether dotted or solid, having two bosses creates numerous problems:

- When customers think they're bosses, they may overstep their authorities and micro-manage internal service providers. This disempowers shared-services leaders who lose a degree of control over their resources, impairing their performance.

- With contention for control of the same resources, it strains the very relationships that it's supposed to strengthen.

- Staff may treat the business unit as a boss to be obeyed rather than a customer to be pleased. This can dilute their entrepreneurial spirit because they feel obliged to follow the boss' orders, rather than proactively bring new ideas to their customers. And they may not take needed time away from the client's projects to invest in their businesses within a business, for example, in training and innovation.

- There's a risk that the shared-services person will speak for clients (as if he/she is a member of the client group) rather than facilitating relationships between clients and everyone in the organization. This undermines customer focus and clients' control of the support services, rather than strengthening it.

- The group with a line to one client may not consider other markets for their services.

  This problem is particularly serious for any Building Block other than Account Sales, which is the only function that should be dedicated to a client. In every other Building Block, a dotted line to one client constrains its ability to serve other clients, and is likely to perpetuate an inappropriate substructure where the wrong Building Blocks are substructured by client (case study in Chapter 4).

- Even in client-dedicated Sales functions, further problems occur when the group serves multiple clients, not just the one holding the dotted line.

  It's inappropriate to expect one client to speak for his/her peers. And it makes no sense to work through one client to build relationships with other clients. Ultimately, the dotted line biases the person to favor one client over the others.

- There's a risk that this person will not adequately serve customers within the organization itself.

- And, of course, it's unfair to put staff in situations where they may get conflicting directions from two bosses.

The Matrix Structure

A matrix structure is similar to dotted lines in its treatment of clients as bosses over shared services staff.

The term "matrix" has come to mean many different things. Here, we'll use its original definition. [43] Staff report to the shared-services organization. But they have two bosses within that organization — the two dimensions of the matrix:

> Boss #1: Staff report to a client-dedicated manager for project direction. He/she sets priorities and directs work. Essentially, they're substructured by client, like decentralization.
>
> Boss #2: Technical direction comes from a profession manager, who's responsible for hiring, developing, and supplying qualified talent to the client-dedicated groups; establishing consistent methods; facilitating the sharing of lessons learned; managing staff's performance; and providing viable career paths.

This is an attempt to get the best of both decentralization and shared services. But does it?

In most professional functions, projects may be short- or long-term. Each project requires a unique mix of skills, not the same team every time. And specialists with scarce competencies may be needed by multiple project teams in parallel.

To maintain a matrix structure and get the right talent on every project, the entire organization would have to be restructured for each new project, an obvious impossibility.

In practice, the matrix structure assigns the same team to a client, no matter what competencies their current projects may require. It's no better than decentralization, except for one thing: Some of the disciplines are better coordinated, and there may be more sharing of methods and discoveries.

A Better Answer

The truth is, dotted lines from shared-services groups up to clients, or the solid lines of a matrix structure, aren't needed. There's no reason good communications can't occur without a reporting line. And it doesn't take a line for staff to know that their performance is, to a great extent, measured by their customers' satisfaction.

Dotted lines and matrix structures represent confusion about the roles of "customer" and "boss." The problems with dual reporting can be avoided, and the same objectives achieved, by giving clients all the respect and authorities they deserve as customers.

If a client requests a dotted line or a dedicated group in a matrix, a good answer is, "Let's talk about *why* you want that." No doubt their objectives can be better achieved through an effective Account Sales function, a customer-focused culture reinforced by performance appraisals that consider customer satisfaction, and resource-governance processes that empower clients to control the priorities of internal service providers.

SYNOPSIS

» Dotted lines and dual reporting in a matrix structure create many serious problems.

» Generally, the motive for these structural compromises can be addressed better by ensuring that clients are respected and empowered as customers, not bosses.

Chapter 23: Plan-Build-Run

Here's an example of an overly simplistic theory of structure that doesn't work well.

Situation

A CIO was dissatisfied with the pace of innovation in his organization. It seemed his staff didn't have time to look into emerging technologies, strategic applications of IT to the business, or new methods of delivery.

This lack of time was a resource-governance problem. But he tried to solve it with structure, by forming a Planning group as part of a "Plan-Build-Run" structure. He noted that others in his industry had similar functions, sometimes calling them "Emerging Technologies," "Innovation," or "Research" groups.

The Planning group, managed by Bonnie, was responsible for client relationships, analysis of requirements, emerging-technology research, business planning, and innovation in all its forms.

The Build group, under Dennis, was responsible for developing the systems that Bonnie planned.

The Run group reporting to Rob was responsible for operations.

Under the CIO were the following groups:

- Bonnie, planning and innovation (plan)
- Dennis, development (build)
- Rob, operations (run)

Analysis

Bonnie's Planning group was supposed to put more emphasis on innovation. But it produced the opposite effect. Here's what you'd see in the Rainbow Analysis:

Rainbows: Bonnie's Planning group was supposed to be aware of the latest developments in all the many domains of information technologies (every branch of Engineering), knowledgable about clients and their business strategies (Sales), and also experts in methods of business planning (Coordinator).

But people can't be experts in more than one thing at a time; and Bonnie couldn't afford to hire an expert in every IT discipline. Instead, she recruited a few very bright (and expensive) people.

They tried to keep up with trends. But there's no way her staff, however bright they may have been, could have kept up with emerging technologies across the entire spectrum of IT. With all the thinking about the future funneled through her small group, Bonnie became a bottleneck for innovation.

Scattered campus: This structure splits the Engineering line of business into "thinking" (Bonnie) and "doing" (Dennis).

Bonnie had authority over systems designs, but not accountability for delivering projects. This induced an "ivory tower" perspective among her staff. They felt free to advocate new technologies without regard for their practicality.

Meanwhile, Dennis' developers were accountable for the delivery of projects, but they didn't have a say over the design of those systems. They were disempowered (the mirror of Bonnie's authority without accountability). Of course, morale suffered.

Feeling like second-class citizens, they resented Bonnie's group, which took away from them all the fun, strategic thinking — the most interesting and motivational challenges. This resentment undermined teamwork.

Because of internal jealousies, strife, and the resulting lack of collaboration, what little research that was done in the Planning group was not always put into practice. At times, Dennis' staff even attempted to prove that Bonnie's group was wrong, to show that the people who disempowered them weren't really needed.

Meanwhile, lots of talent (in Dennis' group) was wasted. Again, innovation was impeded.

And with the learning component removed from their jobs, their skills became obsolete; their productivity and the quality of their work diminished; and their sense of professionalism deteriorated into a production-line atmosphere.

Alternative

In a healthy organization, everybody both delivers today's work and thinks about the future, within their lines of business.

To give everybody time for innovation, address your resource-governance processes, not structure, as described in Chapter 30.

SYNOPSIS

» The "plan-build-run" structure creates a bottleneck for innovation (in the planning group), and disempowered jobs (in the build group).

» Every internal entrepreneur should be responsible for planning and innovating, as well as delivering his/her current products and services.

» Reserve time for innovation through your resource-governance processes, not structure.

Chapter 24:
New Versus Old

This case study is another example of the problems that occur when structure is used to solve resource-governance challenges.

Situation

An apparel manufacturer embarked on a major initiative to pioneer "mass customization" — quickly producing clothing that's customized to buyers' requirements on a mass-production assembly line. *44*

Of course, this business strategy was very dependent on IT. It generated a number of highly-visible projects, revamping core IT applications to accommodate the need for rapid integration of orders into production plans. These huge projects required a mix of skills from throughout the IT organization.

The CIO worried that if he assigned these very strategic projects to the existing applications managers, they'd get lost in the pressures of day-to-day support and maintenance tasks.

He also worried about the converse — some managers might focus on these "hot" projects and neglect less glamorous maintenance work which was essential to keep the current business running.

Therefore, he separated his development staff into two groups. He assigned a young, up-and-coming manager, John, to the new "strategic systems" group, and transferred staff with the needed skills from Jean's existing applications development group.

Under the CIO were two Applications Engineering groups:

- John, strategic systems development
- Jean, legacy systems maintenance

Essentially, applications engineers were substructured by new-versus-old instead of their professional specialties. [45]

Analysis

Here's what you'd see in the Rainbow Analysis:

Scattered campus: When new applications went into production, John's group (who designed the new systems) had to train Jean's group (who maintained them), requiring a dual learning curve. They tried to rotate staff from one group to the other, but the numbers never worked out right; rotations were never sufficient to preclude the need for this redundant training. This added costs and slowed things down.

Inappropriate substructure: Dividing Engineers by new/old, not by type of application, reduced their degree of specialization. The same Engineering specialties were present in both groups. The organization needed two of every type of expert.

For example, in manufacturing systems, instead of having one group focus on scheduling and another on inventory systems, this structure had one developer and one maintainer, each of which had to know both scheduling and inventory systems.

Reduced specialization resulted in higher costs, lower quality, and a slower pace of technical innovation.

Furthermore, John's developers didn't have a strong incentive to "do it right the first time," since most of them didn't have to live

Chapter 24: New Versus Old

with the consequences of their sloppy work. They could get the glory by getting projects out quickly, and leave it to Jean's group to fix their errors.

Even if they cared, John's group didn't get feedback on the quality of their work, since they didn't have to confront maintenance problems. So they had little opportunity to learn from past mistakes and improve their quality.

The resulting lack of quality required expensive repairs once systems were in production. Without first-time quality, costs rose; and in some cases, it disrupted the business.

Furthermore, this structure created a bottleneck that slowed the pace of innovation. John's group did all the future-oriented thinking and learning. This wasted bright minds in Jean's group.

There were also grave impacts on morale. The split created "two classes of citizenship" within the IT organization:
> one building the future, the other maintaining obsolete systems;
>> one politically visible, the other with no way to be heroes but plenty of opportunities to fail if anything goes wrong;
>>> one with great career opportunities, the other in dead-end jobs.

John's staff became arrogant. Morale in Jean's group suffered — rightly so since they found themselves in inferior jobs. Motivation and productivity dropped, and turnover rose.

The internal rivalries undermined teamwork. Of course, this had a cost. The tremendous knowledge of current systems and data in Jean's staff was not well utilized in the design of the new systems. This adversely affected quality, cost, and speed.

Also, costs went up for an interesting reason. Nobody in John's group wanted to be "sent down" to the maintenance group. But these projects couldn't go on forever. Staff spent a lot of time worrying about their next job, and prolonged projects unnecessarily to maintain their privileged position.

At other times, the resource inflexibility worked the other way. Since Jean's staff could not easily be shifted to projects in John's group, they kept busy with minor enhancements, even though their time might have been better spent on higher-payoff new developments.

In addition to all its problems, this approach to new projects is not scalable. Over time, the organization would have to be restructured whenever major initiatives begin or end.

Alternative

The CIO's original concern was a resource-governance problem. This structure fenced off resources (people's time) for major new projects. But as this case illustrates, using structure to resolve resource conflicts has many serious unintended consequences.

In a principle-based structure, every group both delivers its current products and develops the new products that will obsolete its current work. In a healthy organization, everybody has a future.

Resource-governance issues can be addressed in another, more effective way. Budgeting and demand-management processes can allocate resources to supporting past products, delivering new projects, and inventing the next generation of products beyond that. (See Chapter 30.)

Chapter 24: New Versus Old

SYNOPSIS

» Two groups producing essentially the same products and services (one new, the other old) compete in unhealthy ways; it reduces the organization's degree of specialization, increases costs, reduces quality, and slows innovation.

» Ensuring a focus on strategic projects while reserving time for support of existing products is a resource-governance challenge, not a matter for structure.

Chapter 25:
Quick Versus Slow (Bi-modal)

Here's an idea that seems to come up in IT circles every decade or so, each time under a different name. The unhappy results are always the same.

Situation

In IT, traditional business systems require a detailed development process, and carefully planned and controlled change. By contrast, customer-engaging web sites need to be developed and modified quickly. The reality is, IT organizations need to be excellent at both.

One such IT organization split its developers into two groups:

- Bob, rapid applications development
- Larry, traditional applications development

Bob's group addressed urgent, strategic needs with small, quick solutions. Larry concentrated on large development projects, as well as enhancements and repairs to existing large systems.

This concept isn't new. In past decades, CIOs have set up small groups to handle short, high-payoff projects, perhaps labeled the "development center" or the "agile development group." Or most recently, "bi-modal IT."

There were benefits. Bob's group produced many valuable solutions. But ultimately, dividing groups by means (methods) rather than ends (products) generated more problems than benefits.

Analysis

Here's what you'd see in the Rainbow Analysis:

Scattered campus: When multiple groups produce the same products by different means, a line of business is fragmented.

This split limited professional collaboration, and forced redundant learning curves. For example, Bob's team didn't take advantage of the extensive institutional knowledge of data and business processes in Larry's group. They had to relearn the business content before applying their methods, wasting a lot of time.

It also became challenging to get these two classes of systems to work together. Bob's customer-facing web sites needed data from Larry's production systems. But neither group was in a position to design a holistic architecture.

Inappropriate substructure: The basis for substructure (quick versus slow) didn't match what people were supposed to be good at (the applications they engineered).

While differing in their methods, both groups had to replicate essentially the same expertise: knowledge of the applications, data, and processes. Bob siphoned staff away from Larry's group; so each of Larry's people had to cover more ground. Meanwhile, Bob's small group had to cover the gamut of data and processes.

Thus, specialization in their core expertise was reduced. This impacted productivity, quality, and the pace of innovation.

Furthermore, the two groups competed for work, instead of a single group offering clients two alternatives.

They tried to cooperate, but the boundary between "short" and "long" projects was unclear. Bob was limited to projects under a certain number of person-hours. But a given project may fall on either side of that dividing line, depending on something as simple as the choice of platform, the addition of a single feature, or the quality of construction.

This competition confused clients. Bob advised them to take the quick approach; Larry gave them the opposite recommendation. Clients weren't sure whom they could trust.

It also impacted quality. Clients loved Bob's rapid response on highly visible strategic initiatives. And the quality of Larry's work was less apparent to them. So Bob's group won in popularity ratings, and became a "loophole" that allowed clients to violate quality and architectural standards, and build solutions that were difficult to support in the long term.

And finally, this structure created "two classes of citizenship." Bob got the glory; Larry just kept the entire company running. Larry's staff had every right to resent Bob's group which got most of the politically visible, strategic projects. Their morale and productivity deteriorated.

Ultimately, this structure was an "enabler" of the problem. The presence of Bob's group reduced pressure on Larry's mainstream applications developers to be more responsive to the spectrum of large and small projects. They became inflexible in their processes, and unresponsive to urgent projects which still required their meticulous methods.

Also, innovation is discouraged and people cling to old ways when their jobs are defined by existing methods. Neither group had the job of discovering new methods, tools, and processes — new means to the same ends.

Alternative

There are two root causes that need to be addressed, neither of which is structure.

One is methods. Every Engineering group needs to flexibly adapt its methods to match the project at hand. Each should be trained and willing to apply "light" methods to small projects, and more rigorous methods where needed.

Another problem may be a project-definition (requirements planning) method that tends to define *all* a client's possible needs, rather than just a well-focused next step. When many requirements get bundled into one big project, the organization naturally becomes less responsive. This should be addressed with a Sales function trained in a business-driven opportunity-identification method.

The second root cause is resource-governance processes. All available hours may inadvertently be allocated to large projects, those which can be identified during the budget planning process. This leaves no resources available for the quick projects that arise throughout the year.

Instead of using structure to reserve resources for small, strategic projects, a portion of the staff's time (or budget money) should be held aside. This is a challenge for the internal economy, not structure. (See Chapter 30.)

Splitting Bob's group out of Larry's did not address either of these root causes.

The better answer is to use structure to provide a single group of experts in each domain of engineering, and then address the lack of responsiveness by correcting its root causes: methods, and the internal economy.

SYNOPSIS

» Splitting a "quick response" group from the traditional delivery group induces unhealthy internal competition and inappropriate design biases; and it reduces specialization.

» Every line of business should be flexible in its delivery methods, offering both "quick and dirty" solutions and carefully constructed high-quality solutions.

» Reserving time for quick, strategic projects is a matter for methods and resource-governance processes, not structure.

Chapter 26: The Pool

Here's one last case study on the confusion of structure and resource-governance issues.

Situation

An IT department needed to be flexible in assigning staff to a continually changing mix of projects, while still maintaining expertise in each application. Its solution was a "programmer pool."

Carl managed the applications engineers (systems analysts), with groups under him dedicated to particular sets of applications.

His applications managers staffed their project teams by drawing on a pool of programmers (generalists) reporting to Brian.

Under the CIO were the following Engineering groups:

- └ Carl, applications managers
- └ Brian, programmer pool

This put the more experienced people in the key role of applications experts and project leaders, and provided resource flexibility as demand in one area rose while in other areas it diminished.

But the drawbacks were considerable.

Analysis

Here's what you'd see in the Rainbow Analysis:

Inappropriate substructure: The people in Brian's programmer pool could not specialize in any particular applications. They were generalists, expected to work on any system, as the workload demanded.

Thus, Brian's programmers lacked the in-depth knowledge of applications that comes with continuity and specialization. Nonetheless, they made important design decisions as they developed code. Quality suffered.

To make matters worse, there was little incentive for first-time quality. Programmers were not accountable for results, other than just showing up and doing as they were told. And at the end of the project, the programmers would move on to another project, leaving Carl's applications managers to deal with any problems they left behind. As a result of this authority (to write the code) without accountability (for the errors), again quality suffered.

Also, things took longer, since programmers were repeatedly at the bottom of the learning curve, and had to find their way around new applications with each new project.

This structure also created serious motivational issues. Brian's programmers felt like second-class citizens, not valued for their skills. They came to resent the applications engineers (who, admittedly, were a bit smug). Teamwork suffered, as did their effectiveness.

Programmers in the pool were especially demotivated because it was so hard for them to graduate to the status of applications engineers. Every new project took them into a new area of technology; so they didn't have a chance to learn any particular application in depth. They were trapped in dead-end jobs without the ability to gain competencies through specialization that would get them promoted out of the pool.

Alternative

Staffing flexibility was needed, but this was not an effective (or kind) way to attain it.

A better approach is to form technology-aligned Engineering groups comprised of managers, analysts, and programmers (all job levels).

Load balancing can be accomplished by loaning staff between groups when necessary (Chapter 35), while still providing everyone with a permanent home that determines their area of expertise, and a career progression as skills are gained.

SYNOPSIS

» A pool of staff who serve various product-dedicated Engineering managers reduces specialization, and impacts quality and speed.

» It also creates dead-end jobs, and is highly demotivational.

» Flexibility in resources can be accomplished by loaning staff temporarily to peers, while still providing everyone with a permanent home and a specialty.

Chapter 27:
Boundaryless/Network/Cellular/Organic Organization

Beyond cynicism about the relevance of structure, some people are *anti*-structure. They claim that organizational boundaries cause people to focus only on their parochial objectives, undermining teamwork. They cite the "boundaryless organization" as an antidote to this fragmentation. Let's take a look at the consequences.

Situation

The term "boundaryless company" is credited to Jack Welch, then CEO of General Electric. He said, "...boundarylessness is an open, trusting, sharing of ideas. A willingness to listen, debate, and then take the best ideas and get on with it." [46]

Welch describes GE as an environment where "you want to reach across the boundaries separating you from your customers and your suppliers and your colleagues overseas." In GE, Welch claimed that teams combine the right people, regardless of functional area, level in the hierarchy, and geography, to quickly solve problems and meet competitive challenges.

Some have taken this healthy vision to its illogical extreme. They use the term "boundaryless" literally to mean the elimination of boundaries (rather than teamwork across boundaries). [47]

In these flat organizations, there is no hierarchy. People develop their own unique blends of expertise. Small, autonomous groups form dynamically as people join teams based on their interests.

This has been labelled an "organic" organization. [48] Another name is a "cellular" organization. [49] Still another term people use is the "network" organization.

Proponents see the network organization as a replacement for a hierarchically structured organization chart. One even said, "If it can fit on a traditional org chart, it's not a network." [50]

These concepts are a reaction to the faults of hierarchical control. Proponents say that people are more empowered, and teamwork is organic. They tout advantages including flexibility, responsiveness, creativity, continuous innovation, and a high level of empowerment.

A well-known case study is Valve Corporation, maker of online games and a network platform for multi-player games. [51]

Valve's approximately 400 employees can work on whatever they want (be it engineering, marketing, testing, or IT). Teams form around ideas (such as products) as individuals choose what they want to work on. Products die if people lose interest, and evolve if people are excited about them.

Valve has been creative and successful, both in terms of impact on their market niche and profitability. But there is a down-side. Known jokingly as "Valve time," they're notorious for not meeting announced dates for product releases. And some say it's not as great a place to work as they'd have you believe, with an informal hierarchy of power based on personalities rather than structure. [52]

Network organizations have their limits, not the least of which is scalability. For example, GitHub, a software and services provider to software developers, started as a boundaryless organization with its inception in 2008. [53] Hoping to enhance creativity, founders Chris Wanstrath and Tom Preston-Werner asked staff to pursue their own ideas, and contribute to others' ideas which interested them.

But as the company grew, the lack of structure became a constraint. For example, employees didn't know where to take their concerns about problems with colleagues. In 2014, a complaint of gender discrimination revealed a complete lack of HR policies and processes.

Coordination among engineering, legal, marketing, sales, and other departments was critical to growth; but with no department heads, it was left to chance. Staff had difficulty knowing what was expected of them, and struggled to get the support they needed from others with complementary skills.

This limited the company's ability to take on complex projects, increase the frequency of product updates, organize a big developer conference, and secure major partnerships with large firms such as IBM — all activities which require well-orchestrated teamwork among a variety of specialists.

In 2014, six years after its inception, the company had grown to over 600 employees, and abandoned its experiment with a boundaryless organization.

In the extreme, some proponents even go so far as to posit that corporations will be replaced by a network of independent businesses who collaborate to monetize any ideas that pop up within the network, in any way participants can imagine. [54] They see this as an alternative to hierarchical companies focused on narrow business strategies such as industries or segments.

While it's argued that this maximizes innovation, it gives up the economies-of-scale that result when companies focus on a specific niche and penetrate that market to the point of building scale. Ironically, proponents tout the revenue potential of such a network as coming from the many licenses its small companies can sell to traditional large companies. Again, it's not scalable.

Analysis

The Rainbow Analysis explains why vague and fluid domains and accountabilities create these serious problems:

1. Gaps: One purpose of structure is to define staff's specialties. In an amorphous structure, people develop competencies based on personal interests, often "dabbling" in multiple disciplines.

Nobody is responsible for ensuring that new technologies or disciplines are covered somewhere in the organization. Without coordination of domains, there's no guarantee that all the needed skills will be available within the organization (e.g., HR at GitHub).

The resulting gaps lead to unreliable processes, high costs, low quality, and less innovation. And with some specialties altogether missing, lucrative opportunities may fall through the cracks.

2. Rainbows: Staff are accountable for many different Building Blocks. This makes it tough to define people's goals and metrics, and to manage performance.

Thus, an amorphous structure depends on self-starters — people who will take initiatives on their own to develop new skills and product lines. Others, who might be quite competent once channeled by management, are left to flounder. While some people prosper, others become disenfranchised and demotivated, and find the lack of focus in their jobs disconcerting and stressful.

3. Scattered campus: With everybody choosing their own fields of study, there's no guarantee that multiple people won't choose the same specialty (overlaps). This leads to costly redundancies. And when a domain is covered twice, some other domain is not covered as well (reduced specialization).

Fragmented efforts in each line of business lead to fragmented results. While the amorphous organization may respond well in the short term to individual needs, it does poorly at producing a well-integrated product line. The result is higher life-cycle costs, and perhaps lost enterprise synergies.

And without some overriding management structure, it does poorly at aligning resources with enterprise business strategies.

Furthermore, people don't feel "ownership" of a line of business. This inhibits entrepreneurship and motivation. For many, it means just doing as their told, floating from project to project but not thinking creatively about the future of an internal line of business.

In practice, a boundaryless organization does not engender effective teamwork. Although people may be willing to work together, the lack of role clarity makes it difficult to know where to get specific kinds of help. Or a needed specialty may not be available because it didn't happen to interest any of the current staff. Teams are based on social relationships and happenstance, or perhaps the intervention of senior executives, rather than a reliable process that ensures the right talent on each team.

Furthermore, a lack of clear individual accountabilities makes it hard to clearly define authorities. When there is a difference of opinion within a team, the structure provides no basis for problem resolution. The spirit of teamwork can dissolve into battles for control.

An amorphous structure may work well in small groups of high performers, where it's easy to know everybody. But it's not scalable to large organizations where coordination of many distinct disciplines and processes is essential.

Alternative: Hierarchy Isn't "Versus"

Organizational performance begins with having available all the right specialists. A hierarchical organization chart is a very good way of laying out domains — without gaps and overlaps. It allows people to focus on a specialty and own a business within a business.

Then, it's essential to flexibly assemble the right talent on teams to handle each unique challenge, without regard for people's location or level in the organization.

But management hierarchy is a cumbersome way to coordinate work. Traditional top-down control is "bureaucratic." [55] It's slow to respond, limited in its flexibility (its ability to deal with diverse customer needs), and it stifles innovation.

By contrast, the business-within-a-business paradigm and the meta-process for teamwork (described in Part 8) link well-focused specialists across structural boundaries. As people with projects "subcontract" with other groups for help, staff combine dynamically on teams. Processes are tailored to the unique needs of each project or service. Multiple processes run in parallel, each drawing on shared talent.

A principle-based structure is highly organic in its operating model — not through chaos, but rather through well-focused entrepreneurships and explicit, but dynamic, processes of cross-boundary teamwork. [56]

With a principle-based organization, you can "have your cake and eat it too." You can use hierarchy to lay out the organization chart, and then a flexible process of teamwork to coordinate work.

SYNOPSIS

» Unclear boundaries lead to gaps and overlaps, and hence reduced performance.

» Flexible, dynamic teamwork does not result from a lack of structural boundaries. To the contrary, clear boundaries *enable* effective cross-boundary teamwork, telling staff where to go to find the specialists they need on their teams.

» A hierarchical organization chart should be overlaid with a flexible process for forming cross-boundary teams (Part 8). Both are necessary; neither alone is sufficient. And the absence of both is debilitating.

Chapter 28: Leaderless Organization (Sociocracy and Holacracy)

Over the years, there have been numerous attempts to envision a leaderless organization run by a network of peers — not a chaotic boundaryless organization, but one where processes take the place of hierarchy. The latest, and one of the most elaborate, is "Holacracy."

Situation

Holacracy was invented by Brian Robertson in 2007. [57] It was based on Sociocracy, derived from an idealistic political theory of the 19th century and Quaker practices [58], and applied to organizations in the mid-20th century. [59] Holacracy was brought into the limelight when Zappos adopted it in 2013.

Holacracy is a reaction to symptoms of unhealthy organizations such as authoritarian managers, disempowerment, loose delegation that can be overridden, and micro-management.

It takes aim at the power vested in the management hierarchy. Elliot Jacques represents this "old school" of thought where executives do all the strategic thinking, managers make tactical decisions, and everybody else just does as they're told. [60] This, of course, creates bottlenecks, wastes talent, and demotivates staff.

Even in less afflicted organizations, Holacracy cites problems with poorly designed structures like vague roles, periodic major restructurings that are costly and disruptive, and a lack of agility.

Holacracy is a pendulum-swing to the other extreme. Leaders have *far less* power. "Holacracy... removes power from a management hierarchy and distributes it across clear roles, which can then be executed autonomously, without a micro-managing boss." *61*

Holacracy promises flexibility, agility, innovation, and employee engagement and motivation. But does it deliver?

How Holacracy Works

Same Hierarchy, Different Names

Despite the rhetoric and cult-like terminology, Holacracy is very much like a hierarchical organization chart in many ways. The big difference is that it's a hierarchy of roles, not people.

The organization's top executive has the authority to structure the organization, defining "Roles" at the next level. A Role (also called a "Circle" when it includes more than one person) is, essentially, a group with a specific mission.

The top executive recruits leaders into those Roles. These leaders at the next level divide their Role into subordinate Roles (the next tier of structure) and recruit leaders. And so on.

Thus, a hierarchy of Roles (groups) results. Each is led by a manager appointed by the leader above him/her. Each has the authority to decide the structure and recruit staff at the next level.

Employees as Independent Agents

A key difference from traditional organizations is that Holacracy treats employees as independent agents. People are invited to take on Roles, and have the option to accept or decline.

Chapter 28: Leaderless Organization (Sociocracy and Holacracy)

People can take on multiple Roles in multiple groups. Employees decide for themselves the allocation of their time to each group.

Employees don't have a supervisor. Rather, a portion of their work is overseen by the manager of each group they join.

Governance and Tactical Meetings

Holacracy operates through two types of meetings: Governance, and Tactical. It prescribes detailed agendas for these meetings.

Each group's leadership team holds regular (typically monthly) "Governance" meetings to adjust the structure and people's assignments. Boundaries may be changed. New Roles may be defined, or existing Roles eliminated. If a Role becomes too big for one person, it's divided into subordinate groups.

Leaders are instructed to focus not on principle-based, lasting designs, but rather on expedient solutions, since the structure can easily be updated in successive meetings.

Operational decisions within each group are made in regular (typically weekly) "Tactical" meetings that deal with ongoing operations, project execution, program design, customer satisfaction, synchronizing group members' activities, and employee issues.

Analysis

To its credit, Holacracy seeks the kind of empowered, entrepreneurial organization described in this book. In fact, empowerment is mandatory. As a group defines sub-groups, it must delegate authorities along with accountabilities. And no one can disempower others (without their permission).

Also, structural decisions are made through a participative process, as is recommended in Part 9 of this book.

But there are many crucial differences.

No Science of Structure

Although a "Governance Process" is defined in detail, Holacracy doesn't offer insight on how best to define structure. Other than a version of Principle 1, the Golden Rule of empowerment, there are no guidelines for structural decisions (nothing akin to the other Principles). And there's no conceptual framework for defining domains (nothing akin to the Building Blocks).

There's no guarantee that the organization will cultivate expertise in all the specialties it needs, so gaps occur. And groups may be defined in ways that create overlaps (Principle 3) or reduce specialization (Principle 4). It doesn't preclude conflicts of interests (Principle 5). And with specialties scattered among groups, professional synergies are lost (Principle 6).

Doesn't Encourage Specialization

Employees don't have a "home group" which defines their specialty; they can accept multiple Roles in multiple groups based on their interests, drifting from group to group, from specialty to specialty. This leads people to become generalists (Principle 2).

The lack of specialization undermines innovation as well as performance. At Zappos, so far people can't cite any innovations that have improved customer service or increased sales.

Not Based on Business Within a Business

Perhaps most profoundly, there's no guarantee that jobs will be defined as businesses within a business (Principle 7). Roles (groups) are defined by any combination of their purpose (function), "domain" (what it owns and controls), and/or accountabilities (defined as activities, not products and services).

Since groups aren't defined as internal lines of business, no entrepreneur is accountable for offering a comprehensive, innovative catalog of services. No one is accountable for ensuring that internal services are delivered at competitive costs. And no one is accountable for planning the future of each business within a business. The lack of entrepreneurial thinking leaves an organization adrift, and puts its future at risk.

Disempowered Managers

Unfortunately, Holacracy's emphasis on empowerment doesn't include managers, termed "Lead Links."

The role of Lead Link is a bit different from managers in conventional structures. They can recruit staff; allocate resources; establish priorities and strategies; define metrics; and resolve issues that get in the way. But they don't control staff's time; people take on multiple Roles and decide their own priorities.

Thus, a Lead Link is accountable for the performance of the group, but he/she doesn't have the authority to control its resources (staff's time). Here's an example:

Tony Hsieh, CEO of Zappos, wanted someone to implement a time-reporting system so that everyone could see how others are allocating their time. No one volunteered. So Hsieh asked an employee if he would do it. He declined, saying he was too busy. Hsieh pushed until he relented. The only way Hsieh could get the project done was by overriding Holacracy. [62]

Managers are further disempowered when they share with others the job of representing the group at the level above ("Rep Links"). Again, they have accountabilities without matching authorities.

Lack of Performance Management

Holacracy leaves unclear who has the authority to hire people into the organization, decide compensation, evaluate performance, and discipline (or fire) people.

People act is independent agents, moving fluidly among Roles in various Circles. The Roles have a boss; people do not.

So no one has the authority to deal with performance problems. A Lead Link (manager) can fire someone from a Role; but not from the company. And the more people spread themselves among groups, the harder it is to appraise their performance.

Thus, people aren't held accountable for delivering their commitments, or for their productivity, quality, effectiveness, or innovation. There's significant risk in expecting everybody to perform well without oversight, just because they're empowered.

Here's a sad case example of the disempowerment of management: [63]

At a small education company, Betsy had always been a problem. She was emotional, angry, and difficult to talk to. She verbally punished people for disagreeing with her. She turned business debates into personal animosities. She aired differences in public,

Chapter 28: Leaderless Organization (Sociocracy and Holacracy) 219

and rallied allies to help her fight with her peers. She blamed any attempts to address her problems on sexism. She turned the group against Larry, the group's Lead Link, using Holacracy's anti-authority ethos. Her negative role model and behaviors affected the whole team; morale plummeted.

In this Holacracy, Larry was held responsible for the group's performance. But there was nothing he could do to manage Betsy.

Betsy ultimately moved on to another group, and caused the same turmoil there. But for Larry, the damage was done. This talented young leader was forced out of the company.

Another perverse effect of the lack of performance management is that people who are good at studying the rules and playing the game succeed; while better performers and perhaps the more creative souls are lost in the chaos.

Compensation Is Not Based on Performance

Holacracy also makes compensation decisions more difficult, since there are no job descriptions or performance metrics.

Holacracy users are forced to invent new compensation schemes. For example, at Zappos, a "badge" system pays employees more for moving into different Roles and gaining competencies in other professions, or in the theory of Holacracy. The system is not based on contributions, and seems to encourage dabbling in many things rather than specializing in one. [64]

Weak Cross-functional Teamwork

Holacracy does not define mechanisms of cross-boundary teamwork. Instead, each Role is expected to be self-sufficient. It must recruit staff with all the specialties it needs. Holacracy forms a new Role to address every new initiative.

Thus, it's difficult to coordinate the delivery of large projects, for example, major strategic initiatives which engage many functions and people throughout the enterprise.

Its Own Form of Bureaucracy

Although it purports to eliminate bureaucracy, Holacracy devotes a great deal of people's time to building consensus on every governance and operational issue. And it's fraught with complex language, and time-consuming procedures and meetings.

Getting the job done takes a back seat to all these procedures. Everyone is accountable for *improving* the way they do their work (resolving "Tensions"); *defining* projects and "Next-Actions" (tasks); *tracking* their own work (and making status available to all); and *doing* the work, as time permits — in that order.

The structure is in constant flux. Each new challenge may require a structural change. And everybody in the enterprise is expected to keep up with those changes.

Holacracy may support flexible reassignment of talent among teams, but at a cost of significant amounts of everybody's time spent on organizational issues (some say more than half a day each week per person).

Alternative

Many of the values at the core of Holacracy are embodied in the Principles in this book — empowerment, entrepreneurship, clear roles and boundaries, and participative decision making.

But Holacracy makes a critical error: **It confuses groups in a structure with cross-boundary teams.** This is why each new initiative requires a reorganization.

A better answer is found in a principle-based structure. It creates empowered, entrepreneurial groups, and gives everybody a stable home that allows them to specialize — the very reason organizations exist.

Then, overlaying the organization chart, a team-formation process (described in Part 8) draws talent from these groups, bringing together just the right people for each unique project and service.

With effective teamwork, there's no need for reorganizations every time a new project arises, or when groups need help from other specialists. The organization is both stable and dynamic.

A well-designed organization has none of the ills that Holacracy rails against, and delivers all the benefits (and many more).

SYNOPSIS

» Holacracy is a reaction to authoritarian managers, disempowerment, loose delegation that can be overridden, and micro-management.

» Staff are independent contractors, working where they please.

» Holacracy lacks mechanisms of cross-boundary teamwork, and hence forms new groups whenever new teams are needed.

» It leads to reduced specialization, gaps, overlaps, lack of entrepreneurship, disempowered managers, weak performance management, and poor cross-boundary teamwork.

» A healthy organization is made of empowered, entrepreneurial groups, and processes of cross-boundary teamwork that assemble the right specialists on teams from throughout a stable structure.

~ PART 6 ~

Design: How to Assemble the Building Blocks into an Organization Chart

"Engineering is the conscious application of science to the problems of economic production."

Halbert P. Gillette [65]

The same science that allows you to see problems in any organization chart (the Rainbow Analysis) can be used to design an optimal structure for your organization.

You can tailor a structure to your unique needs by assembling the Building Blocks in your own way. You might combine some small Building Blocks into a single group. You might split apart a single Building Block into multiple groups. You might even ignore some Building Blocks that don't apply to your organization.

Just by using the Building Blocks as the basis for your design, you'll satisfy many of the Principles. And by applying the Principles as you assemble the Building Blocks, you'll avoid the temptation to accommodate personality issues, individual career aspirations, fads, and political pressures.

Using the science of structure, you'll end up with a highly effective organization that's tailored to your unique needs. This Part explain how to do that.

Chapter 29:
Clean Sheet Versus Tweaks

When planning a structural change, the first choice you have to make is between a "clean sheet of paper" approach and a few small "tweaks" to the existing structure.

The Clean Sheet approach uses your current structure only to catalog all the lines of business that exist today, and hence are needed in the new structure. Beyond that, it ignores the current organization chart and starts fresh, designing an optimal new structure that includes all those functions (and perhaps more).

The Tweaks approach begins with your current organization chart, and moves a few specific domains from group to group.

One thing you might want to know as you consider this choice: **Many small tweaks are actually more painful to implement, and far less effective, than a Clean Sheet approach.**

The pain is the result of human nature, our innate territorialism. If a manager's group is going to stay essentially intact, it's emotionally difficult to give up pieces of it. People naturally try to protect their territories. But with a clean sheet of paper, no manager has an extant territory to defend.

The pain is far worse if the tweaks are not implemented all at once. A series of small changes is more disruptive and scary to staff than a single, well-communicated big change. It's like picking at a scab. The wound stays open. Staff are held in a continuing state of uncertainty and fear, wondering when the next shoe will drop.

Furthermore, the organization remains in confusion and less than fully productive for a longer period of time.

As for effectiveness, if many changes are needed, a whole-system perspective is required. It's difficult to design a structure as an integrated system and get everything in the right place with a series of small tweaks. The Clean Sheet approach is far more effective.

Also, tweaks are far more difficult if you have many "rainbow groups" such that their accountabilities have to be sorted into multiple domains. It's even worse if you have "rainbow people." As groups (and individuals) give up and take on functions, there's confusion and a risk of gridlock. Many rainbows suggest the Clean Sheet approach.

But if the current organization chart is close to right and only a few small adjustments are required, then the Tweaks approach is the right answer for you.

"The incremental approach to change is effective when what you want is more of what you've already got."

Richard Pascale [66]

The Rainbow Analysis (Chapter 17) is a very effective way to decide which approach is best for you. It gives you a list of problems, and helps you estimate the magnitude of the changes needed in your organization.

Even if you're only making a few small tweaks, the science of structure can help you design incremental changes that solve pressing near-term problems, and are in keeping with a consistent evolutionary direction. This way, each small change brings your organization closer to the ideal without introducing new and different problems.

The remainder of this Part describes the Clean Sheet design process. But all the guidelines can equally be applied to designing incremental changes.

SYNOPSIS

» The Clean Sheet approach uses your current structure only to catalog all the lines of business that exist today, and then starts fresh, designing an optimal new structure that includes all those functions (and perhaps more).

» The Tweaks approach begins with your current organization chart, and moves a few specific domains from group to group.

» Clean Sheet is actually easier and more effective than many Tweaks. Tweak what you have only if changes are limited to a few small adjustments.

Chapter 30: Assumptions

Let's start with a few assumptions, without which one might inadvertently compromise the Principles.

It may be tempting to use structure to address problems whose root causes are elsewhere. But if you don't understand and fix the true root causes of your concerns, you won't get a fully satisfactory solution. Meanwhile, you'll sacrifice the effectiveness of your structure, and set yourself up for another restructuring in the future.

By assuming you'll solve other problems in other ways, you'll isolate and highlight, rather than mask, those other issues. And you'll be free to design an optimal structure.

Competent Leaders

First, assume that your leaders are competent. Never compromise the structure to avoid dealing with individual performance issues.

Individual performance problems may be the result of a good person in a bad job (e.g., impossible requisite variety, conflicts of interests, disempowerment). In this case, structure will solve the problem.

But if you have a low-performing individual, the right answer is to find a job in which he/she can succeed, or to release him/her.

The same is true of high-performers. Don't design jobs just to boost their careers. This may help the individuals; but a structure that depends on someone's unique talents isn't stable. When the individual leaves, it's difficult to replace him/her, creating a new performance issue. You may even have to restructure again.

It's always best to design a structure without thinking about which individuals will occupy the boxes, and deal with individual performance issues through individual performance management.

Solve Resource-governance Problems with your Internal Economy

Resource-governance processes (your "internal economy") are as important as structure in getting an organization to perform well. And they, too, can be designed using a well-documented science. [67]

Clients' control of priorities: For internal service providers, clients want to control priorities (what they "buy" from the organization, whether or not money changes hands).

To give them this control, some leaders dedicate groups to business units. This communicates to those clients the finite resources they have at their disposal, and it makes it easy for clients to control the priorities of their dedicated staff.

But this client-centric structure (akin to decentralization) creates many serious problems, as described in numerous case studies.

And it doesn't solve the problem of priorities in the support groups behind those front-line client-dedicated groups. These support functions become bottlenecks; and clients' priorities may not be respected as each support group decides its own priorities. So this really isn't an effective way to give clients control of priorities.

Giving clients control of priorities is a challenge for your resource-governance processes, not structure. Clients can be given control of spending power rather than dedicated staff, either through chargebacks or (perhaps more practically) through a claim on the organization's direct budget.

Money, not staff, can be divided among your clients. Clients then understand the limits of their checkbooks (your resources), and decide what checks they write (priorities, a.k.a., demand management).

This frees you to optimize the effectiveness of your structure. And clients get a better deal; they can use their spending power to buy any specialists your organization has to offer, not just a group of generalists dedicated to them.

Time for innovation: Another resource-governance challenge is setting aside time for innovation. As the "Plan-Build-Run" structure described in Chapter 23 illustrates, it's unwise and ineffective to use structure to fence off resources for innovation.

It's far better to address the real root causes of the problem in your internal economy:

- **Demand management:** Staff may be so busy responding to clients' unbridled demands that they have no time left for thinking about the future. The solution: Force clients to limit their demand to available resources.

- **Billable-time ratio:** In calculating what an organization will deliver for a given level of budget, people may be expected to "bill" (i.e., dedicate to projects and services) too high a percentage of their time, leaving little time for innovation. The solution: Set aside an appropriate amount of unbillable time for innovation and other critical sustenance tasks before promising deliverables to clients.

- **Time management:** People may not utilize the time allotted for planning, professional development, and other innovation functions. Instead, they may let it be absorbed by day-to-day tasks. The solution: Train staff in time management.
- **Venture funding:** Funding (money and time) for major innovation projects may be lacking. The solution: Set aside a pool of money and hours not allocated to clients' projects, but available for innovation projects, funded by explicit line-items in your budget.

Once you implement effective resource-governance processes, every group can find time for innovation within its own domain. There's no reason to design a structure that separates learning from doing, creates a bottleneck for innovation, and disempowers most of the staff.

Your internal economy is also the way to address many other issues, like giving clients control over their costs (not allocations); responding quickly to small, strategic projects; and providing explicit funding for enterprise-good services.

Trust in Teamwork

People who are resigned to their organization's current, limited abilities to team across boundaries build "silo" organizations which are relatively self-sufficient.

There is some theoretical support for this point of view. Anthropologist Robin Dunbar hypothesized that people can only comfortably maintain 150 stable relationships. [68]

Based on the belief that teamwork is the result of friendships and a common boss who coordinates work, some have applied Dunbar's finding to structure. They divide organizations into self-sufficient groups of no more than 150 people. Dunbar himself said, "The reasons why we might want to think of units of 150 in terms of organisational structures is... so the smallest unit... can stand on its own two feet." [69]

The resulting silos sacrifice performance. They scatter campuses, cause rainbows, and use inappropriate bases for substructure — with all the problems those structural faults entail.

Note that friendships aren't essential to collaboration. Look to the real-world economy for examples. Collaboration extends globally, engaging hundreds of millions of people. Teamwork doesn't depend on personal relationships, or a boss who knows everybody and personally coordinates their activities. Instead, market economics coordinates people.

Like market economics, the team-forming processes described in Part 8 can coordinate teamwork across boundaries in very large organizations, without a common boss or personal relationships.

As to Dunbar's number, small groups should indeed be tight-knit teams with strong social relationships. But this doesn't mean that an organization as a whole is constrained to a size in which staff have personal friendships with everyone they work with.

The purpose of an organization chart is to create loci of expertise. Then, projects cut across structure — drawing the right experts from various groups for each project. When this team-forming process works well, there's no need to create independent silos. Indeed, there are many reasons not to.

So don't constrain your thinking to your organization's current teamwork abilities. Instead, invest in cross-boundary teamwork processes as a part of implementing a new structure. Then, you're free to optimize your organization chart rather than sacrifice performance just to avoid the need for teamwork.

Structure for the Future, Not Just Today

There may be a tendency to catalog the work you do today, and divide that among the boxes on a new organization chart.

It's true that accountability for everything you do today must be placed somewhere in your new structure. But limiting your vision to today's work is shortsighted. You're setting yourself up for another restructuring when things change.

Worse, if discovering future lines of business isn't built into everyone's jobs, it becomes the job of the top executive to plan everybody's future. He/she becomes a bottleneck for innovation.

Instead, it's best to design a structure that will accommodate everything your organization might do, now and in the future.

That doesn't mean you have to hire people to fulfill functions for which the need hasn't yet arisen. It just means that accountability for those future functions must be assigned somewhere in the new structure. For lines of business that aren't needed today, someone should have the job of monitoring developments in the field; sharing with peers what's possible so that opportunities aren't overlooked; bringing in qualified experts (vendors and consultants) as needed while retaining accountability for delivery; and proposing headcount when the time is right.

Designing a structure for the future creates a self-adaptive organization. It means your organization will not only survive the future without major restructurings; it will proactively create that future.

They *Will* Let You

I've heard leaders say that they're constrained in their choices by the expectations of their clients, their superiors, or even their staff.

Internal service providers may feel they have no choice but to dedicate groups to business units because client executives want that. Or they may feel compelled to place a function at the highest level (even though it fits better lower in the structure) purely for political reasons.

Or executives may feel pressured to combine two functions (e.g., Base Engineering and Asset-based Service Providers) because the current manufacturing (infrastructure) manager demands control of the engineering function as a boss, not as a customer.

Don't sacrifice your principles to appease others. Your organization's performance will suffer.

It's better to stand up for what's right and face the resistance now, than to face the blame later for the problems that compromises will bring. Wouldn't you rather spend your "credibility chips" on what's right, than spend them defensively later?

Trust that you can explain the logic of, and how things will work in, a well-designed structure. You'll either convince people, or gain enough of the benefit of the doubt to give you time to prove that healthy organizations are in everybody's best interests.

Conclusion

With these assumptions, you're free to design a structure that's optimal for your organization. The rest of this Part describes a step-by-step process for designing a really great organization chart, customized to your needs.

This design method becomes the core of the implementation process described in Part 9.

SYNOPSIS

» Don't design your structure to accommodate the individual capabilities and career needs of your current leaders.

» To give yourself the freedom to optimize your structure, presume that you'll resolve resource-governance issues in your internal economy, not structure.

» You can design a structure that optimizes specialization if you trust that you'll also build effective cross-boundary teamwork processes (Part 8).

» Structure for the future, not just for today's workload.

» Sell what's right; don't give in to pressures to compromise your design.

Chapter 31:
All the Lines of Business

Now you're ready to begin the process of designing a new organization chart.

You may do these steps on your own. Or much better, you may engage your leadership team in the design process (as described in Part 9). Either way, the steps are the same.

This Chapter, and the remainder of this Part, describe the design method. Part 7 complements this process by exploring a number of special cases and design issues.

Part 8 turns our attention from the organization chart to the processes of teamwork.

Then, Part 9 wraps an implementation process around this design method and the teamwork processes described in Part 8, as well as the challenges of change management.

Start With the Building Blocks

The first step in designing a structure is to list all the lines of business that must exist somewhere in your organization. The Building Blocks provide a framework for this.

Under each of the top-level Building Blocks, list the detailed lines of business that need to be somewhere in your organization chart. As long as your list describes what they produce (their products and services), these detailed lines of business can be very granular and yet still be whole, empowered entrepreneurships.

When we do this as a participative process, we write each detailed line of business on a large post-it note, [70] color-coded to indicate its high-level Building Block as per the Rainbow Analysis. They're placed on flip charts which represent the Building Blocks.

No Gaps

To ensure effective performance in all its missions, your list must be comprehensive. It must pinpoint accountability for every line of business within the scope of the organization, now and in the foreseeable future. (As discussed in Chapter 30, you can structure for the future without having to hire staff in every future domain.)

A comprehensive structure is also essential to be truly business driven, not a "solution in search of a problem."

The Rainbow Analysis of your current structure provides you with a list of all the lines of business that exist today. Add to that any missing Building Blocks or emerging lines of business that you'll want to include in the new structure.

As you brainstorm all the lines of business, don't forget those which serve customers within the organization itself, like Sales, Coordinators, Base Engineers, and People-based Service Providers.

Other Enterprise Services

If you're an organization within an enterprise, you consume support services provided by other shared-services departments. For example, every department needs services such as those provided by Corporate Human Resources and Finance.

You have two choices:

Model A: Everybody in your organization works directly with that shared-services department. It's trusted to satisfy all your needs, and there's no need for a point of contact and coordination within your organization.

Model B: A group within your organization serves your staff. It uses the shared-services department as much as possible (as an extension to its staff, just as it would external vendors), and finds another way to satisfy any needs they don't fulfill.

Figure 24: Relationship with Shared Services (Model A/B)

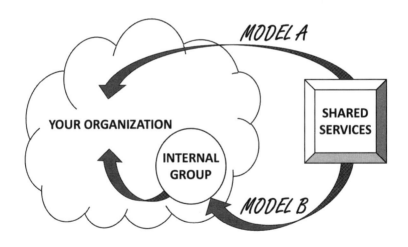

Model B is decentralization of that other support function. So from an enterprise perspective, it's not ideal.

But you have to look after the effectiveness of your own organization. So if there's any doubt about the ability of that shared-services department to serve you, Model B is safer. It doesn't replicate anything provided by the shared-services department (since your group "outsources" to them whenever possible). But it ensures that all your needs are met.

If you choose Model B, add that line of business (post-it) to your list, as a People-based Service Provider.

Results of this Step

At this point, you should have a comprehensive list of all the lines of business needed within your organization, sorted by Building Block. If you've written them on post-its, they're placed on flip-charts representing the Building Blocks.

SYNOPSIS

» First, gather a list of all the lines of business needed in the new structure.

» Sort them by the Building Blocks.

Chapter 32: Clustering the Lines of Business

The next step is to move the lines of business (those post-it notes) into clusters that will become groups in your structure.

In other words, a box on your organization chart will be defined as a collection of these lines of business.

Creating Legitimate Rainbows

When you keep the lines of business within each Building Block together, you've satisfied all the Principles.

But in your unique situation, there may be good reasons to move a specific line of business (a post-it note) from one cluster and to another, creating small "rainbows." The synergies with another Building Block may be stronger than the synergies with other lines of business within the same Building Block.

For example, it's common to place a project-management office (PMO, a People-based Service Provider) in the Coordinators cluster alongside the Organizational Effectiveness line of business. There are synergies. Both focus on *how* we work. And structure (OE) drives the layout of accountabilities in a project plan (PMO).

You'll just have to be careful that one mission doesn't overshadow the other. For example, routine PMO services may keep staff so busy that there's no time to lead transformations (OE).

Also, you'll have to be careful that the ethos of one doesn't bias the other. For example, despite the Coordinators' focus on consensus, a PMO doesn't need to be concerned if different kinds of projects or different managers utilize different project-management methods. As long as all the projects succeed and comply with all the rules (including status reporting and documentation), commonizing methods may do more harm than good.

There are other cases where the problems caused by a rainbow can be more difficult to manage. So watch out for "bad" reasons to move a line of business to a different cluster, including these:

- **They work together:** If you're committed to cross-boundary teamwork, there's no need to move a line of business into another cluster just because they frequently work together.

- **Accountability:** You can hold managers accountable even if they use other internal groups as suppliers (just as they retain accountability when they utilize vendors). They're just as accountable in the role of customers as they are as bosses.

- **Headcount:** In small organizations, efficiency is particularly critical. But remember, a clean structure is more efficient than rainbow groups created under the guise of "not enough headcount to separate functions." And you'll have the option to combine small clusters in the next step.

Be conservative about creating rainbows. Be sure there's a good reason, and that you can manage the risks that these rainbows create.

And when you do move a line of business to a different cluster, keep your language clear. The line of business doesn't magically become a different Building Block. You've made a conscious decision to create a rainbow. So don't pretend that the line of business has changed colors.

Breaking Up a Line of Business

You might feel the need to break up a line of business and put pieces of it in multiple clusters (a scattered campus). This always has consequences. Be very careful to assure yourself that the benefits outweigh the costs.

Geography is rarely a reason. (See Chapter 36 for a discussion of remote locations.)

Generally, breaking up a line of business is only worthwhile when you're less interested in the benefits of specialization, professional coordination, synergies, and career paths; and more interested in each group's ability to act independently.

When you do break up a line of business into multiple groups, be mindful about the basis for substructure. Domains should never be separated by tasks, processes, means rather than ends, or subsets of the services that a line of business offers. And remember, every line of business is responsible for both today's deliverables and for its future products and services. Never split today's work from future work in the same domain.

The entire line of business should be replicated in each cluster; and their domains should be differentiated in some other way, e.g., by the customers they serve.

Of course, if you do break up a line of business, you'll need to set up ongoing mechanisms to monitor the decision and deal with any resulting problems, such as:

- **Scattered campus:** You may build special forums to encourage collaboration, i.e., "professional interests groups." As sub-specialties arise in one group and not others, you may ask the experts to cross-train rather than leave other groups to re-learn. But don't expect one group to serve as a "center of excellence" for others (as discussed in Chapter 21).

You may also have to keep an eye on conflicts of interests, and ensure that no gaps or further overlaps in domains arise.

- **Basis for substructure:** When domains are divided on any basis other than its bottom-of-the-T, there's a loss of specialization. To account for this, you may recognize a stretch assignment (too much for one person or group to do well) in people's goals and metrics. And you may use external vendors and consultants, shared among the groups, when a higher degree of specialization is needed.

  If you're big enough to afford a central group of specialists to back up the scattered groups, then perhaps those groups are really "field technicians" who work under the direction of the experts, a People-based Service Provider. (See Chapter 36.)

If multiple instances of a line of business are necessary, add them to more than one cluster on the list. In a workshop, make copies of the post-it, and place them in different clusters.

Chapter 32: Clustering the Lines of Business

SYNOPSIS

» Cluster lines of business by Building Block as much as possible.

» Avoid Rainbows and Scattered Campuses unless there's a good reason which overrides the inevitable problems.

» Be prepared to manage the consequences of any rainbows or scattered campuses you create.

Chapter 33:
From Clusters to Boxes

At this point, you have clusters that are, for the most part, a single Building Block. But you've selectively created some rainbows, and perhaps you've had to split a line of business among multiple clusters.

The next step is to convert these clusters into groups, that is, boxes on your organization chart.

Some of your clusters may be huge, while others may be tiny. So you may want to divide the big clusters into multiple groups, and combine some small clusters into a single group. The salient issue is span of control.

Span of Control

"Span of control" means the number of groups (or individuals) who directly report to a manager.

Anthropologists tell us that people tend to organize themselves into hierarchies. Small groups form, ranging up to around 15 people. Three to four of these groups are assembled under a boss. And three to four of those bosses are put under a boss at the next level up, and so on. [71]

That suggests that there's a natural tendency toward a span of three to four (at levels above the most basic group) — at least in the hunter-gatherer societies that were studied.

But that may not apply to your organization. This ancient span may be because hierarchy was the way they coordinated activities across groups. But we have alternative coordinating mechanisms, e.g., plans, policies, resource-governance processes, and the team-forming process described in Part 8.

So in a modern organization, what's the optimum span of control?

It's not simply a matter of headcount. In cybernetic terms, it's a matter of *variety* — complexity, volume (headcount), and pace. [72]

There's no one right span for all the different Building Blocks, since they differ in the variety they process. Each of your clusters must be examined individually.

Calculating Span: Working Managers

Span of control is generally equal to the number of people at the next tier who report to a manager (direct reports). But for very small groups, there's an exception.

Imagine that there are only two people in a line of business. But it's a distinct line of business, and deserves the focused attention of its own group.

One of those two people could be appointed the manager. But this person certainly will not spend all his/her time supervising the other person (a span of one)! He/she will be a "working manager" who will both supervise the other person and deliver some of the group's work.

As the number of subordinates grows, the time spent delivering work diminishes until the manager becomes a full-time supervisor.

Without this concept of working managers, you may be tempted to combine small clusters unnecessarily. Doing so doesn't really save any money; the same amount of work — supervision and delivery — needs to get done; it's just distributed differently.

If you're challenged with a target span of control (insensitive to each manager's variety), remember to count working managers as fractional supervisors, with the remainder of their time treated as a subordinate in that group. For example, in a group of two, the manager may spend just 20 percent of his/her time supervising a group of 1.8 headcount (the other person plus 80 percent of himself/herself), a span of 9.

Appropriate Span of Control

The clusters will become "tier-one" leadership positions reporting to the top executive (who is "tier zero"). So the first issue is the executive's span — the number of groups that directly report to him/her.

- Too many clusters (too wide a span), and the executive won't have time to manage so many people while still fulfilling his/her diplomatic, strategic, and leadership functions.

    If this is the case, you'll have to combine some clusters to reduce the executive's span.

    (Alternatively, a "deputy" may expand the variety-processing capabilities of the executive. [73])

- Too few clusters (too narrow a span) is costly. Jobs at the next tier may be too big and difficult to fill. It may force another supervisory layer further down in the structure. It distances the executive from the staff. And it may mean that you've inadvertently combined Building Blocks in ways that create rainbows.

If this is the case, you'll want to divide some clusters.

At tier one, leaders are typically seasoned executives who are, to a great extent, self directed. With competent leaders managing well-focused and empowered groups, a reasonable span for the top executive typically ranges from 8 to 14.

At lower tiers, it depends on the function:

- In relatively stable, routine lines of business (like Service Providers), a wide span (8 to 12) is feasible. At the level of first-line managers (below which are individual contributors), groups may be as large as 12 to 15 staff.

- On the other extreme, where there's lots of complexity and things are changing quickly (perhaps Base Engineering), a reasonable span may be 5 to 8. And first-line managers may have groups of just 6 or fewer staff.

These are only rough guidelines. For each cluster and each tier, you'll have to decide the appropriate span for your organization.

Too Wide a Span, Combine Clusters

Greater span is better, up to the limits of requisite variety. It reduces span for the next tier, making those jobs easier to fill. It might avoid an additional supervisory layer at lower tiers ("flattening" the structure). And with fewer layers, it brings all the staff closer to the executive.

However, there is a maximum span of control, beyond which a manager's job is too big to be feasible. If there are too many clusters under the top executive, you'll have to combine clusters to reduce the executive's span.

There's no need to combine clusters just because a cluster appears too small. Don't worry about trying to make the jobs equal in scope. It's okay to have managers of differing job grades at the same tier. And a very small cluster could be led by a working manager (see above) or a self-managed group (Chapter 34).

Furthermore, there's no need to combine clusters just because you don't have enough people to fill all the jobs at this level. You can keep the boxes on the organization chart separate, and simply put the same manager's name in more than one box. This keeps the lines of business clear, and provides a scalable structure as the organization grows. And who knows? You may find a talented individual at lower levels who's ready for a (albeit small) leadership position.

The only reason to combine clusters is to reduce the span of the level above.

If you need to combine clusters, to the extent possible, combine lines of business vertically within the same Building Block, not horizontally across Building Blocks. Even if there aren't natural synergies in these combinations, at least this avoids conflicts of interests.

But if you have to combine across Building Blocks, beware creating rainbows with difficult conflicts of interests. Some bad combinations to avoid are:

- Engineers with Service Providers (innovation with operations).

- Sales with Engineers or Service Providers (unbiased business advice with bias toward a subset of the product line).

- Audit with anything else. (Audit must be arm's length, and judging others compromises one's ability to partner with them.)

Too Narrow a Span, Dividing Big Clusters

If there are too few clusters, the executive's span is too narrow and the jobs at the next tier may be too big, requiring a tier-one manager to cover an unreasonable intellectual territory (impossible requisite variety). It may force an additional supervisory layer at a lower tier. Also, it risks losing focus on distinct lines of business. It may even create undesirable rainbows and conflicts of interests.

To widen the executive's span, divide the post-it notes into two or more separate clusters. Choose which lines of business to keep together based on similarities in their professions and products/services.

Here are specific guidelines by Building Block:

Engineers

If the Engineers are too big a cluster, the first split should separate Applications from Base Engineers.

Applications Engineering can be subdivided by the purpose of the applications. For example, in IT, applications developers (along with the applications they support) should be divided by data objects (topics).

Base Engineers can be subdivided by their technologies or disciplines. For example, in IT, Base Engineering might be divided into platforms, end-user computing, and software engineering tools and methods.

Service Providers

The first split should separate Asset-based (infrastructure services) from People-Based Service Providers.

Asset-based Service Providers are experts in operating assets that produce services. The right basis for substructure is by the types of services they produce (not the types of assets). There may be layers of services (where one service is built on the foundation of lower-level services). [74] When this is the case, it's best to keep the layers together as much as possible.

People-based Service Providers are also experts in services. Divide them by type of service. Start by separating services that are product oriented (such as customer service and training) from more general "business office" support function.

Coordinators

The first split should separate business-oriented from technical Coordinators. Beyond that, split Coordinators by the topics they coordinate.

Sales and Marketing

Account Sales for companies should be dedicated to specific large customers or industries (not by geography). For internal service providers, Account Sales should by divided by the clients' organization chart.

Retail Sales, on the other hand, is appropriately substructured by geography. It may be clustered under the same manager as Account Sales, or in the same cluster as Marketing.

Chapter 33: From Clusters to Boxes 251

Function Sales domains are defined by a client profession or business-process. It too may be clustered with Account Sales (combining all the Sales functions), or with Marketing (combining all the services that support Account Sales).

Marketing is first split into Marketing Communications and Market Research. At the next level, these can be split by the types of services they offer. For example, Marketing Communications might be divided into planning and branding, channels, and support services.

Audit

Audit is best subdivided by what they examine and judge, e.g., the types of policies and standards they examine (like finance versus regulatory compliance).

Good Reasons for the Wrong Basis for Substructure

As per Principle 4, the ideal way to subdivide a line of business is by its sub-specialties (described above). Any other basis for substructure reduces specialization.

But in some situations, other factors make it worthwhile to sacrifice some degree of specialization. Here's a case example: [75]

A distributor of grocery products sells to micro-store owners (as well as large grocery chains). Its CEO knew that more face-time with customers would increase sales. But unlike professional buyers in the large chains, these "mom-and-pop" store owners were only willing to spend so much time and money with each sales representative, no matter how broad their product lines.

While the "depth of the T" in knowing customers is important in sales, in this case, face-time was more important. Realizing this, he divided the sales force by category (akin to aisles in the grocery store) within each geographic territory,

Small store owners then allocated an amount of time to each sales representative; and with multiple sales representatives calling on each store owner, the company in total got more face-time from each store owner, and sales went up.

Of course, this increased costs. The additional face-time per customer required a bigger sales force. However, the additional revenues more than compensated for higher selling costs.

There was another cost. This structure made it more difficult for sales representatives to build relationships with store owners. But that was already impeded by the limited time that store owners were willing to spend with each representative. So not much was lost.

When you divide a cluster by any basis other than its sub-specialties, be mindful of the consequences and look for other ways to manage the risks.

Lower Tiers

Organizations over approximately 100 people require another layer of managers at tier two. Above a few hundred, you'll need tier-three managers as well. In very large organizations, there may be many supervisory layers.

This means dividing tier-one clusters (now leadership positions) into tier-two clusters (management positions at the next tier), and then those into tier-three clusters (and so on).

Apply the same guidelines for span of control, and for dividing lines of business (post-its) into clusters (jobs) at the next tier.

That's Your Organization Chart

Part 7 discusses a number of special situations and design options that may help you split or combine clusters.

Once this step is done, you have clusters of lines of business that represent management positions at each tier. You're ready to draw your new organization chart.

SYNOPSIS

- » Turn clusters of lines of business into jobs (boxes on your organization chart) by considering span of control (how many direct reports each leader can manage).

- » Span of control is a function of variety — complexity, volume, and speed.

- » Account for working managers who both supervise others (part time) and do the work of the group (with the rest of their time).

- » Avoid creating conflicts of interests when combining clusters by combining vertically within a Building Block, not horizontally across them.

- » Use the appropriate basis for substructure when dividing clusters.

~ PART 7 ~

Special Situations and Design Guidelines

"You gotta do what you gotta do."

Anonymous

This Part describes a number of common challenges that, on the surface, might seem to force you to deviate from your ideal structure. But in this Part, you'll see ways to approach these challenges without compromising the Principles.

The special situations discussed are:

- Self-managed groups
- Shared people
- Remote locations
- Project management office
- Compliance and governance

Chapter 34:
Self-managed Groups

What if a line of business is small, perhaps just a few people, and no one person in it is significantly more senior than the others? An example is a small group of "gurus" in a distinct discipline, such as a few experts in a technology.

You could put the staff under the manager of a related line of business. But that may risk extending that manager too far, or creating an undesirable "rainbow."

An alternative is to leave the line of business as a box on the organization chart, but with no manager. This is termed a "self-managed group." The group manages itself as if there were a supervisor over it.

In these situations, a self-managed group has advantages: It maintains a focus on that line of business, rather than burying it within another line of business. And it provides a growth path for a function that may someday warrant a manager.

Self-managed groups may also be useful when members of the group need to be placed at a higher level in the structure in order to get the political visibility that they need to be effective. For example, in an internal service provider, a few senior Account Sales staff may all report directly to the organization's executive in order to attract high-level talent to this critical function and to make them visible to internal client executives.

Self-managed groups are also helpful when members of the group need to be placed at a higher level in the structure in order to get the job-grade (compensation) they need, such as when antiquated

HR policies do not permit people to report to someone of the same grade level.

Of course, the people and the line of business still need supervision. In a self-managed group, **staff share the duties that a boss otherwise would perform**. How they do so should be agreed by the members of the group and the manager above them.

The duties of supervisors (at any level) are listed in Appendix 2. For each, the members of a self-managed group must agree on how these tasks will be performed and how decisions will be made. Choices include the following:

- The boss (a level up) does it.
- A "lead" (perhaps on a rotating basis) does it.
- Any individual can do it independently.
- The group collaborates on it, e.g., through consensus.

This sharing of supervisory duties should be reinforced with some degree of shared destiny to ensure collaboration. For example, some portion of everyone's performance appraisal should be based on the performance of the whole group.

Contrary to first impressions, self-managed groups do not save headcount. Supervisory duties must still be fulfilled, even if they're shared by the members of the group. The total supervisory workload may actually increase as a result of the transactions costs of rotating or sharing supervisory duties.

The disadvantages of self-managed groups emanate from their weaker leadership. They may not be as coherent in their business strategies and directions, nor as strong in representing their views at the next level up (e.g., competing for resources and influence).

They also may be slower to acquire new methods and tools. Thus, new lines of business tend to get off to a slower start.

Self-managed groups also require more attention from the manager above them, at a minimum to do individual performance management, coaching, and career counseling. And the manager above them must ensure that the supervisory duties are being performed effectively by the group.

Thus, while self-managed groups are supposed to behave as a single direct report, they inevitably reduce the feasible span of the manager at the next level up, which might cost an additional layer of structure elsewhere in the organization.

Self-managed groups are not a goal in themselves (although empowerment is). They're simply a way of treating situations that warrant a structural separation but not a manager. The approach should be used with caution; and the proper supervisory agreements and rewards must be cultivated to make them work.

SYNOPSIS

» Self-managed groups aren't a goal in themselves. They're just a way to deal with special situations where a distinct line of business is fulfilled by a small group of equals.

» Self-managed groups must agree on how they'll fulfill all the duties of a supervisor (Appendix 2).

» Self-managed groups reduce the feasible span of the manager at the next level up.

Chapter 35: Shared People: Temporary Duty

Sometimes it's appropriate for one manager to loan an individual to another, part time or full time, temporarily or indefinitely. Borrowing a term from the military, I call this "temporary duty" (even though it may be an ongoing arrangement).

Temporary duty may be needed because certain individuals have skills that are beyond the domain of their group ("rainbow" people). In such a case, it's not good to expand their group's domain to encompass their skills, for three reasons:

- This would create a "rainbow" group, and perhaps overlapping domains.

- It would create an anomalous structure that may outlive the individual's tenure, causing difficulties for others who follow in that position (or create the need for another restructuring when the individual moves on).

- It would set a precedent that it's acceptable to violate the Principles.

Instead, it's better to put that person on loan to the other group whose domain includes that person's other skills. That's temporary duty.

Temporary duty may also be used to move portions of headcount from one group to another as workloads shift. This "load balancing" is particularly handy in smaller organizations.

Borrowing a person is not a contract between the two groups for the delivery of a service, since the lending group cannot sell services outside its domain. It's simply the loan of a person. All accountability for the work remains with the receiving group, which manages the borrowed individual as if he/she were a part-time employee. Receiving managers may even pay the individual's salary for that period of time.

If the situation is permanent, temporary duty is equivalent to the individual holding two part-time jobs, reporting to two different managers who individually supervise portions of his/her time.

But it's not "dual reporting" where two managers supervise a person doing a single job, as discussed in Chapter 22. The person works in two distinct domains, and hence has two distinct part-time jobs.

Regardless of the longevity of the arrangement, the two managers must be very clear about the allocation of the person's time, and must agree on when that time is available to each. For example, they may agree on a 50/50 split (2.5 days per week to each group); but one group may need the person full-time for one week, in trade for no time another week.

In all cases, the individual should have a "home" group, reporting to one manager who is administratively accountable for him/her. This administrative manager is the point of contact for human-resources issues.

While this administrative manager takes lead responsibility for writing the individual's performance appraisal, the other manager who borrowed the person should contribute to the performance appraisal in proportion to the amount of time spent working under his/her direction.

Chapter 35: Shared People: Temporary Duty

SYNOPSIS

» "Temporary duty" is when one manager loans an individual to another, part time or full time, temporarily or indefinitely.

» Temporary duty accommodates people with multiple skills, and load balancing as demand shifts, without compromising the clarity of domains.

» The two managers must be clear about the sharing arrangement, and have the authorities and accountabilities as if each has hired a part-time employee.

Chapter 36:
Remote Locations

In geographically dispersed organizations, some remote locations may have very few staff. But the organization still has to deliver all its products and services to clients in that location.

There are four ways to address remote locations.

Rainbow Groups

One solution is to create a "rainbow" group that combines all the Building Blocks. This is similar to decentralization, although the shared-services executive still has supervisory authority.

Its advantages are its simplicity, since minimal teamwork across geographic distances is required, and everyone reports to one boss.

However, this approach has all the disadvantages of a silo.

First and foremost, it reduces specialization, and may create jobs in remote locations that are too broad for anyone to succeed at.

Furthermore, remote people will not be part of the core professional team, so they'll have limited input to enterprise directions.

Explicit mechanisms must be established to coordinate research, product offerings, engineering standards, professional practices, etc. Despite these efforts, remote people rarely receive sufficient technical direction from the appropriate experts (who are busy with their own projects); so consistency is at risk. Dis-integration of the organization's product line — due to rogue groups doing their

Chapter 36: Remote Locations

own thing — increases support costs and undermines enterprise synergies.

Finally, establishing "rainbow" groups sends a signal that it's acceptable to violate the Principles. It sets a bad precedent.

Generally, location-specific "rainbow" groups should be avoided. They're only necessary when collaboration between the remote location and headquarters is so difficult that the location must be run autonomously, despite the costs and risks. With the internet and modern collaborative tools, this is rare in most businesses.

Assuming your managers can supervise people who are geographically dispersed, other alternatives are generally more attractive.

Site Coordinators

In larger remote groups where the headcount is sufficient to permit individuals to specialize in a single function, each specialist can report to the appropriate manager at the central location.

This has the advantages of a single boss, consistency in technical directions, and remote staff participation in professional directions.

In this alternative, one of the remote people (typically the most senior) should be appointed "site coordinator" to deal with location-specific issues such as personnel policies, dress codes, and hours of operation. Site-oriented issues should be clearly defined, to delineate the responsibilities of site coordinators from those of managers. They're generally issues that affect all employees at a location.

Additionally, site coordinators can serve as an agent for the remote staff's real managers. The site coordinator may help distant managers with the generic (not domain-specific) supervisory duties listed in Appendix 2. The manager for each remote individual should clearly define which of these duties he/she wishes to delegate to the site coordinator.

Site coordinators should have input to the remote staff's performance appraisals (which are done by the real manager).

Site coordinator duties are distinct from a client-liaison function, i.e., Retail Sales for that location (although the Retail Sales person may happen to be designated the site coordinator).

Site-coordinator duties are over-and-above this person's normal functional responsibilities, though of course the additional workload must be taken into account.

Technician Services

In small remote groups where staff work in multiple domains, the remote group may be deemed a "Technician Services" business (a People-based Service Provider). Technicians sell their time to others within the organization, acting as their agents ("eyes and hands in the field"). All technical direction comes from the subject-matter experts who hire them.

Technicians don't sell anything directly to clients. They communicate directly with clients; but they're just agents of another group, and accountability flows through the appropriate function.

Technicians may report to a supervisor in the same location (if it's big enough), or to a manager who supervises technicians in multiple locations.

This solution doesn't work for Account Sales or Coordinators, where delegation to field technicians (which requires clear documentation of what they're to do) isn't generally feasible. These functions should be delivered centrally, even to remote locations.

Temporary Duty

If there's a predominant function in a location and all the staff there are involved in that to some degree, the manager and staff may report to that function. To fulfill other functions in that location, any of the staff (manager included) can be loaned to other groups as temporary duty (Chapter 35).

Trade-offs

While rainbow groups are best avoided, the other options are each viable in the right circumstances. Figure 25 summarizes their pros and cons.

Figure 25: Alternatives for Remote Locations

	Cost of teamwork	Single boss	Consistency	Input to specialty	Cross-training
Rainbow group	Low	Y	N	N	Y
Site coordinators	High	Y (distant)	Y	Y	N
Technician svcs.	High	Y	Y	N	Y
Temporary duty	High	N (distant)	Y	Y	N

SYNOPSIS

» Small groups of staff in remote locations don't have to be treated as "rainbow" groups or independent full-service organizations.

» In some cases, each person can report to the right central group, and one of them can be appointed "site coordinator" to deal with site-specific supervisory issues.

» In some cases, the staff at the remote location can be treated as "field technicians" — generalists who act under the direction of central groups.

» In some cases, staff report to a single central group, and some are loaned to other groups as temporary duty.

Chapter 37: Project Management Office

Excellence in project management is essential to reliable project delivery. On large, complex projects, it's particularly critical. What's the best way to incorporate a PMO in your structure?

Traditionally, a project leader is accountable for managing every aspect of the project, including the tasks assigned to every team member. But for large projects, few people have sufficient project-management skills. Some leaders address this by creating a small group of "super project managers" for the difficult projects. Here's an example of the problems that creates.

Case Study: The PMO that Manages Projects

Since Fred's applications engineers were weak in project-management skills, the CIO asked Kathy to form a Project Management Office (PMO).

Kathy wasn't there to help Fred manage his projects. It was the other way around. Kathy was accountable for big or complex projects, and Fred was just a "body shop" supplying people to work on Kathy's projects. This led to numerous problems.

Kathy's project managers were experts in project management, not applications engineering. Nonetheless, they controlled who was on the project team and their assignments, schedules, and key design decisions that affected delivery dates. With non-technical people calling the shots, quality suffered.

Exacerbating this, the incentives were biased against quality.

Kathy wasn't accountable for long-term applications support; but she was certainly accountable for on-time delivery. So she cut corners to get projects out on time.

Meanwhile, Fred was disempowered. He couldn't control key design decisions; but he was still accountable for future support, maintenance, operational efficiency, integrations, etc.

Also, Fred didn't really own his line of business. With Kathy taking over delivery of his products from time to time, he wasn't an empowered entrepreneur who could plan its future, define its products, optimize its methods and costs, and evolve its capabilities.

Fred's staff, being just a body shop on all the interesting projects, were demoralized. And they had little incentive to improve their project-management skills — the root cause of the problem.

As you'd expect, Kathy and Fred were at odds, not because they didn't get along but because the structure set them up to fight.

Project Management as a Service

Clearly, project-management experts are important. A PMO in itself isn't a problem. In fact, it's a great way to bring discipline and proficiency to project delivery.

The problems come from a PMO that's *accountable for projects*, occasionally taking over the delivery of other groups' products.

An effective PMO is a People-based Service Provider, accountable for delivering project planning and facilitation services. It helps everyone succeed at their respective projects, without assuming project leadership or accountability for results.

SYNOPSIS

» A project-management office is a People-based Service Provider, accountable for delivering project planning and facilitation services.

» It helps everyone succeed at their respective projects, without assuming project leadership or accountability for results.

Chapter 38:
Compliance and Governance

"Compliance" processes ensure that organizations follow rules, including their own policies and standards, contractual obligations, and myriad laws and regulations. How should compliance be represented in your organization chart?

Compliance is implemented through "**governance**," which means **all the processes that coordinate and control an organization's resources and actions.**

Governance isn't limited to oversight. Controls can be imbedded in the organizational ecosystem — structure, culture, resource-management processes, and metrics. But too often, executives assume that "governance" means a person or committee with the authority to control others. Let's examine this assumption....

(While this case study focuses on regulatory compliance, its lessons equally apply to security, business continuity, quality, safety, and other compliance challenges.)

Case Study: Chief Compliance Officer Accountable for Compliance

Allison was appointed Chief Compliance Officer in the IT department of a huge financial services company. She enthusiastically told me of the importance of compliance in their industry, and hence the stature of her position as the *one person accountable for the compliance of the entire IT function*. "Our CIO gave me the authority to make that happen," she proudly asserted.

Time and time again, history has proven that this approach doesn't work. Here's why:

Others have businesses to run, and they're not going to let a peer get in their way. Sure, they'll comply when it's easy or when they really have to — with the big, visible initiatives. But on a day-to-day basis, Allison had three factors working against her:

- Allison is accountable for compliance, not her peers. They won't put much effort into something that's not in their own performance objectives.

- Others are accountable for business results, and they're not going to compromise their missions to help Allison with her objectives. In fact, they may have incentives to thwart her if compliance gets in their way.

- The third factor is the killer. Others don't need to worry about compliance. It's Allison's problem. So if they mess up and bad things happen, she'll take the blame. Allison may as well have been given the title, "Chief Scapegoat."

Basically, all these factors encourage Allison's peers to find ways around her controls to get their jobs done. Ultimately, they'll succeed.

Who Decides the Trade-offs?

After I explained this, Allison objected, "But what if one guy takes a risk and something bad happens? Then the whole organization suffers the consequences."

"Allison, would you advocate zero risk?" I asked.

"In public, I might have to say yes," she said. "But I know that would be unrealistic. Zero risk would force us to shut down the business, at least for a while. Obviously, we can't do that."

Idealists may claim that compliance helps achieve business results. But realists know that there are trade-offs. To illustrate this, consider the extreme: If compliance means shutting down the business for a while, maybe the right answer is to wait to implement controls, and hope that nothing bad happens in the meantime.

On the other hand, if the risks of non-compliance are huge (such as people getting hurt or very large fines), a rational person would choose to shut down the business to implement controls.

Allison agreed that trade-offs had to be made with a full understanding of both risks and business impacts.

"So," I said, "somebody has to decide these trade-offs. No matter who that is, if something bad happens, the whole organization suffers the consequences. So the only question is, who should make the decisions — you, or the managers running the business?"

Either Allison could study the business and make the decisions, or she could teach others the risks and let those who know the business decide.

Allison honestly felt she was in the best position to decide the trade-offs. "They always sacrifice compliance for near-term business results. I represent the compliance perspective, so I'd make it a priority. I'd take far less risk than they would."

Now we have a battle brewing. Allison is fighting to minimize risks, while managers fight to maintain operations. Are the best decisions really going to come from internal fighting? Unlikely. It would be far more effective if the decisions were collaborative.

Chapter 38: Compliance and Governance

Everyone Must Be Accountable

There's only one way to make managers *want* to collaborate with Allison. "What if they're all held accountable for their own compliance?" I asked.

She had to grant that this would swing the balance somewhat. But she still wasn't satisfied that they'd make the right decisions.

"You know there's a good chance you'll make the wrong decision too, Allison. Given your position, you'll opt for more compliance than they would, and in doing so you might sacrifice critical business results — maybe without even knowing it since you're not in the trenches delivering services."

I reminded her of the Golden Rule: authority and accountability must match. If she makes the decisions, then she has to be held accountable not just for compliance but for everybody's business results. Otherwise, what's to stop her from deciding in favor of too much compliance, sacrificing business results, and letting others take the blame when critical services fail?

"I can't be held accountable for everything going on in the whole organization!" she cried.

"Exactly," I said. "Therefore, you can't be given the authority to make these decisions."

The right approach is to hold *everybody* accountable for their own behaviors, including for their own compliance. Then, they'll willingly implement compliance initiatives to protect their own hides. Overall, the success rate of compliance initiatives is higher, not lower, when authority and accountability are in the right place.

"Okay," Allison sighed, "I'll grant that if they were really accountable for compliance, we could let them decide the trade-offs. But how do we get them to take that accountability seriously?"

"Your first job," I said, "is getting your boss to put compliance in everybody's performance objectives, and to measure them on it."

Compliance as a Service

When everybody is accountable for their own compliance, the organization doesn't need someone who forces others to comply. But it still needs Allison.

The job of a Compliance Officer is to help others succeed with their compliance accountabilities. It's a *service* based on expertise in how regulations affect the organization and its clients.

"But," Allison said, "regulators require a single point of contact. And even if it weren't required, compliance processes cross organizational boundaries. Someone has to look after the big picture."

Allison was right. Compliance is a Coordinator function. It helps others succeed at their accountabilities, including helping them agree on shared decisions and processes.

If regulators request information or impose an audit, she can coordinate the organization's response and serve as the communications channel to the regulator (that single point of contact). But she's just accountable for those coordination services, while everybody is accountable for their own portion of the response.

She also can help individual managers put together their own policies and plans. At a higher level, she can bring stakeholders to consensus on shared policies and plans, and consolidate their individual plans into an integrated organizational plan.

Similarly, Allison can help others agree on shared initiatives that improve compliance. Then, she can help them implement the agreed changes, not as the manager who's accountable for results but rather as a facilitator and subject-matter expert.

As a Coordinator, Allison can manage tests of plans, while everybody remains accountable for their own groups' responses.

Through all her services, Allison can teach others the regulatory requirements, the risks of non-compliance, and the kinds of changes required to mitigate those risks. Educating others equips them to better decide the trade-offs.

Oversight

"One final concern," Allison said. "What if they just don't do anything about compliance? Who's to catch them? Their bosses may not know enough about the regulations to know that they've got a problem."

There may be a need for oversight, I granted. But it shouldn't be mixed with her service role.

"Remember," I replied, "the real auditors are outside — the regulators, hackers (in the case of security), or Mother Nature (for business continuity). If *you* are seen as an auditor, doors will close as you approach, and you won't have much impact.

Remaining service oriented, Allison can sell "compliance assessment studies" that help managers get ready for the real external audit (or know how their subordinates are doing).

"Describing it this way keeps you on their side of the table," I explained, "there to help them, not judge them. You've got to maintain good relationships to implement meaningful change."

The Scapegoat Trap

Many functions can fall into the scapegoat trap by claiming authority over, and hence accepting accountability for, others' behaviors. Chapter 9 described the case of a Safety group that thought it was accountable for safety. Others who might make this same mistake include:

- A "security" group that thinks it's accountable for security, rather than helping everyone to operate in a secure manner.

- A "business continuity" group that unilaterally designs the plan.

- A "quality assurance" or "testing" group that tries to take accountability for quality through inspection and control, rather than by providing a testing service to others who are accountable for producing quality products.

These are all examples of a familiar theme: Total Quality Management. [76] Quality, in all its forms, is an *attribute* of a product or service, not a separate deliverable. Producing products, and the quality of those products, are not two distinct jobs. Experiences in every industry prove the same principle: Responsibility for compliance, safety, security, and every other aspect of quality should never be separated from responsibility for doing the work.

In every case, better results are achieved when everybody is held accountable for their own behaviors. Coordinators help others with their accountabilities. And if oversight (Audit) is needed, it must be kept arm's-length from service-oriented Coordinators.

SYNOPSIS

» Everybody must be held accountable for their own behaviors, including their compliance.

» A compliance function doesn't disempower others by assuming authority over their compliance; nor does it serve as scapegoats with accountability for others' behaviors.

» A compliance function provides Coordinator services to help others succeed with their compliance accountabilities.

~ PART 8 ~

Workflows: You Can't Specialize If You Can't Team

> *Commerce links all mankind in one common brotherhood of mutual dependence and interests.*
>
> James A. Garfield

To this point, we've focused on the design of an organization chart. Now it's time to turn our attention to the processes of cross-boundary teamwork.

Teamwork is not a luxury; it's an essential aspect of organizational design. As per Principle 2, you can't specialize if you can't team.

Indeed, the word "structure" includes both organization charts and teamwork processes. Henry Mintzberg, author and professor of organizational behavior, said it well:

The structure of an organization can be defined simply as the sum total of the ways in which its labor is divided into distinct tasks and... [how] coordination is achieved among these tasks. [77]

Teamwork is often the gating factor in organizational design. Organizations which aren't very good at it deliberately create "silo" groups which can function relatively independently.

Of course, in doing so, they scatter a profession among all the groups that need it, and hence reduce specialization. In other words, they consciously give up some organizational performance to avoid having to invest in more effective teamwork processes.

Even if self-sufficient silos aren't created deliberately, they'll occur automatically if you neglect teamwork processes. No matter what your organization chart says, if you can't get help from your peers, you're forced to replicate their skills and muddle through on your own. Without effective teamwork, groups inevitably evolve into silos of self-sufficient generalists.

The bottom-line is this: The better an organization is at cross-boundary teamwork, the more it can afford to specialize and, in doing so, optimize its performance. So if you want to build a high-performance structure, you must be prepared to invest in the mechanisms of teamwork.

By investing in teamwork, I don't mean team-building. As you'll see in the next Chapter, there's a lot more to it than liking and trusting one another. Effective cross-boundary teamwork depends on explicit mechanisms for forming and coordinating teams.

This Part describes a powerful, flexible, entrepreneurial approach to cross-boundary teamwork.

But first, let's examine some traditional, but less effective, approaches to teamwork.

Chapter 39: What *Not* to Do to Improve Teamwork

When Randy was hired as CIO, he inherited seasoned executives who were technically qualified, generally liked by their staff, and nice people. One little detail: They didn't team at all well.

Sure, they were cordial with one another and cooperated on organizational decisions like policies and plans. But each ran his/her own group as an independent silo with minimal collaboration on projects and services.

Obviously this was costing the company money. Many skills were replicated across groups, and people were spread thinly as they tried to do too many things, which hurt their productivity.

Another problem was that people managed functions they didn't know much about. Applications developers ran their own development servers (without much security, continuity planning, or even back-ups). Meanwhile, the infrastructure group (adept at running servers) had its own applications developers for its billing system.

Each department had its own support functions, like budgeting and purchasing. There were two infrastructure control centers: one for computing, and another for the network. There were even three different help desks: one for PCs and infrastructure, another for applications, and still another for their telecommunications networks. How were you supposed to know which to call when your PC couldn't access an application via the network!?

For lack of teamwork, hand-offs were rough. When an application was ready for production, there were often delays due to poor coordination between the developers and the infrastructure staff.

Perhaps most embarrassing, each department had its own client liaisons (Sales). The organization looked foolish when clients got different, sometimes conflicting, answers from different "single points of contact," and when one hand didn't know what the other hand was doing.

Just one month after taking the job, Randy was mortified when clients were exposed to internal finger-pointing after a project missed its deadline. This was the last straw. Randy took on teamwork as one of his primary challenges.

Team-building

First, Randy hired a consultant to do a one-day seminar on teamwork. The consultant taught leaders the importance of teaming, team problem-solving techniques, and effective communications skills. Everyone agreed with everything he said. But back on the job, nothing changed.

Randy brought the consultant back to do a more extensive team-building process. This time, the leaders spent three days at a resort playing games that required mutual trust, talking about how they felt about one another and their interdependencies, and playing golf. They got to know one another a lot better, and everyone had a great time. But back on the job, nothing changed.

Job Rotations

Getting a bit frustrated, Randy tried some job swapping. The head of the infrastructure group was assigned to applications development. The head of applications went to client support. And the head of client support moved over to the infrastructure group.

Leaders developed a lot of empathy for one another; but still nothing changed. It wasn't long before the former head of infrastructure, now leading the applications group, was requesting money to upgrade his development servers. Meanwhile, the new head of infrastructure fiercely defended his need for his own billing application developers.

This scheme lasted about three months, at which time Randy put everyone back — just moments before the organization collapsed due to leaders who knew nothing about the functions they managed. Cynics among the staff got a good laugh at this experiment.

Process Engineering

Randy's next attempt was to define cross-boundary processes so that people would know their duties within teams.

For service-management, he hired a consultant to train his staff and implement ITIL "best practices" processes.

For applications development, he hired another consultant to implement "DevOps," including a well-defined process (and tools to support it) to move code from development into production.

For other processes, he chartered teams to document and redesign workflows that spanned the organization, supported by another consultant who specialized in Lean-Six Sigma process engineering.

The processes defined everybody's tasks; but no one was accountable for results — the delivery of services to customers. So Randy designated "process owners" to oversee the processes end-to-end.

This introduced a lot of strife into the organization, since these process owners violated the Golden Rule by trying to tell others how to do their jobs. (See the case study in Chapter 3.)

Many months (and dollars) later, few processes were actually implemented. And for those which were, additional headcount was required to administer the processes and coordinate the hand-offs. The organization became more bureaucratic, less flexible, and less accountable.

This was going in the wrong direction!

Forced Consolidation

At this point, Randy gave up on his participative style and mandated a reorganization. He consolidated the three help desks, moved the development server into the computer center, and set up a single Client Liaison group reporting directly to him.

Of course, this resulted in utter chaos for a while. But after a few months, things settled down....

The unified help desk did a fine job of transferring calls to the three "level-two" help desks.

The development servers were moved into the computer center, but applications developers insisted on managing them themselves. So reliability and security issues remained.

And the new Client Liaison group did a great job of coordinating communications with the three departmental "entry points."

Can you see through the smoke? Nothing changed.

Chapter 39: What *Not* to Do to Improve Teamwork

What Was Wrong?

What made it so tough for Randy to develop effective cross-boundary teamwork?

Randy's frustrations stemmed from a simple mistake: He treated symptoms, not root causes. And as we so often find in life, short-cuts cost more than they're worth.

His department leaders weren't stupid or malicious people, and it's not that they didn't get along with one another. Team-building isn't a bad thing; it was just a solution to the wrong problem.

Rigid, predefined processes weren't the answer either.

The problem was that this organization was missing **explicit mechanisms for forming and coordinating teams**.

A metaphor illustrates the importance of explicit mechanisms of coordination:

In New Jersey, there's no general rule about who has the right of way in a traffic circle; instead, the law says that right-of-ways are determined by "historically established traffic flow patterns." The habits of local residents dictate the rules of the road, and the custom is not always posted. The result: In addition to an extremely high accident rate, gridlock occurs when drivers entering the circle take the right of way, forcing those already in the circle to stop and back up traffic such that no one can get out.

In this case, the missing coordinating mechanism caused real trouble.

The next Chapter sets the stakes, defining what "high-performance teamwork" means.

Then, in Chapter 41, we'll examines traditional mechanisms of teamwork and why they don't meet that high standard.

The final two Chapters in this Part describe a highly effective approach to coordinating cross-boundary teamwork, a key element of a principle-based structure.

SYNOPSIS

» Team-building, job rotations, forced consolidations, and engineered processes may not address the root causes of poor teamwork.

» Explicit mechanisms for forming and coordinating teams are required.

Chapter 40: High-performance Teamwork

Most organizations are not simple assembly lines, producing just a few outputs through highly structured, routine processes. The challenge for most organizations is far more complex:

Organizations must produce a diversity of products and services. They run many projects in parallel, each requiring a unique mix of skills and sequence of tasks; and yet all share a set of resources. The results they produce must be tailored to customers' needs, and yet integrated.

This requires a high level of cross-boundary teamwork. So before we get into the mechanics, let's consider the goal: What is "high-performance teamwork"?

First, teams must form quickly.

Second, an organization must flexibly "mix and match" specialists from throughout the organization on teams, based on the unique requirements of each project or service.

Third, complex workflows must be meticulously orchestrated, with clear individual accountabilities that add up to the end result.

And finally, effective collaboration depends on teams that can resolve their differences without a lot of management intervention.

In other words, high-performance teams are both *self-forming* and *self-managing*.

"**Self-forming**" means three things:

- **Lateral:** An organization can't afford to wait for executives to form teams. Even if time permitted, executives can't possibly know enough to get the right people on every team. Self-forming teams come together laterally, with people soliciting help directly from their peers.

- **Just the right people:** For every project or service, a team draws together just the right people from throughout the entire organization, without regard for structural boundaries.

- **At just the right time:** People aren't assigned to teams for the life of a project; they arrive as needed. And when they're done, they leave the team and move on to other things.

"**Self-managing**" means three things:

- **Clear individual accountabilities:** "All for one and one for all" sounds nice, and it's great when teammates help one another. But in teams of specialists with different competencies, it's essential to clearly define individual accountabilities for specific results.

- **Clear process:** The various results that individuals produce must come together to produce each unique product or service.

- **Clear chain of command:** Differences of opinion should be resolved within the team.

    Consider the example of building you a house. What if everyone working on the project (the carpenter, electrician, plumber, etc.) thought he/she worked directly for you? Each might have a different impression of what you want, and collaboration would be strained as they fight for control. And when things don't come together properly, it's unclear whom you'd hold accountable. But when you buy a house from a general contractor, all the subcontractors understand that they take direction from the general contractor.

Similarly, within project teams, trouble would result if everyone felt equally responsible for the end result. When differences arise, they'd be settled by fights, political ploys, appeals to authority, or drawing customers into internal team disputes. This rarely leads to optimal decisions, and risks embarrassing the organization in front of its clients.

Within teams, a clear chain of command should make it clear who has the authority to make decisions.

In the rest of the Part, we'll explore mechanisms to facilitate this high level of cross-boundary teamwork.

SYNOPSIS

» High-performance teamwork means that teams are self-forming: Everybody takes the initiative to form teams by soliciting help from peers. Teams combine just the right skills, at just the right time (not a fixed team for an entire project).

» High-performance teamwork also means that teams are self-managing: Individual accountabilities are clear. The process is clearly defined, and yet tailored to the needs of each project or service. And there's a clear chain of command within teams to resolve differences.

Chapter 41:
The Limits of Traditional Approaches to Teamwork

There are various ways to assemble people onto teams, define their respective accountabilities, and ensure that their contributions come together to produce the intended result.

Henry Mintzberg described five choices: [78]

- **Direct supervision:** A person (such as a common boss) has authority over the work of others, instructs them in a coordinated way, and monitors their results.

- **Standardization of skills:** Training in a discipline or profession ensures coordination of work.

- **Standardization of outputs:** Well-defined results coordinate everyone's efforts; e.g., specifications ensure that a part made by one division fits the machine made by another.

- **Standardization of processes:** Procedures and flow-charts coordinate activities (process engineering).

- **Mutual adjustment:** Processes of lateral communication are used to coordinate work.

This Chapter examines the limits of the first four approaches, and sets the stage for a powerful approach to teamwork based on mutual adjustment (described in the next two Chapters).

Direct Supervision

Direct supervision is a simple approach to teamwork. People who need to work together are put under a common boss, who assigns them to teams, defines their tasks, and directs their activities.

This traditional approach to teamwork is based on the assumption that people will only work together if their boss tells them to. It has severe limits. Here's a case example: [79]

In manufacturing, the production line has to be scheduled based on demand; and just the right raw materials have to show up at just the right time.

In a rapidly growing food company, this wasn't working well. Production managers did their job. Supply-chain managers did theirs. But their schedules weren't meshing.

To get them to work together, the Chief Operating Officer appointed a "facilities manager" for each of their three production locations. Reporting to each was a production manager and a supply-chain manager. The job of the facilities managers was to ensure that scheduling and production worked well together.

Things did work better. But while solving one problem, this structure created many more.

The company was no longer able to manage its production capacity globally. One plant could be overwhelmed while another worked below capacity.

And best-practices in one facility weren't applied to the others.

Furthermore, the company lost global buying power.

All three facilities reported to a common boss, the COO. So why didn't he coordinate collaboration across facilities better? There was no way that COO had the time to manage that level of detail on a daily basis.

By depending on direct supervision to create teamwork, this company gave up the synergies across manufacturing facilities that were critical to competitive advantage.

While direct supervision is the most widely practiced approach to teamwork, it's the least effective.

Expecting senior managers to form project teams creates a bottleneck that constrains responsiveness and flexibility.

Worse, using the management hierarchy to coordinate teamwork creates independent silos. This causes so many problems: rainbows, scattered campuses, and inappropriate substructure. The results are higher costs and reduced performance due to lack of specialization, lost synergies due to fragmentation, and limited career paths.

We have created trouble for ourselves in organizations by confusing control with order.

Margaret J. Wheatley [80]

Standardization of Skills

Another coordinating mechanism is standardization of skills, where each member of a team understands the skills and activities of every other member of the team well enough to anticipate their actions and adapt to them.

Chapter 41: The Limits of Traditional Approaches to Teamwork

An example is a hospital operating room, where a surgeon, an anesthesiologist, and a nurse work well together even though they don't report to a common boss.

In this approach, people's actions can be anticipated just by knowing something about the position they hold and the situation the team is in.

As well as it works in some settings, standardization of skills has limited applicability to professional work. Managers and professionals don't do the same procedures day after day. So one cannot anticipate another's actions simply by knowing the other's profession.

Standardization of Outputs

Standardization of outputs means specifying the interface between the work of two groups so precisely that they don't need to coordinate their work much further (other than quantities and times).

For example, when an automobile manufacturer orders sub-assemblies from its suppliers, precise specifications are provided. Thereafter, they only need to coordinate quantities and schedules, not the details of the work.

Standardization of outputs only applies to copies of a work product, not to the collaboration required to initially design the specifications (i.e., not to unit #1). It works very well in mass-production situations, where the effort put into defining precise specifications pays off over the entire production run.

However, for custom work such as that of most professionals, each project has unique requirements. Every project is unit #1. Standardization of outputs doesn't apply.

Standardization of Processes (Process Engineering)

Standardization of processes coordinates teams by pre-defining the flow of work (the process) and everyone's tasks within it.

One well-known form of this was business process reengineering (BPR). Popular in the 1990s, it promised greater efficiency, obliteration of red tape, and streamlined workflows. [81]

A more modern name for this is Lean. Like BPR, Lean is a method by which teams design more efficient processes. It's often combined with Six Sigma, a set of quantitative methods for measuring processes, plus a process-design method that utilizes those metrics.

All these popular process-engineering methods are founded on the original "socio-technical systems analysis" which dates back to the 1940s. [82] They all bring teams together to streamline processes, deal with variability (quality issues), and define people's tasks.

In some cases, processes are similar (at least at a high level) in all companies, such that common processes can be defined. In IT, an example of pre-defined, generic "best practices" is the Information Technology Infrastructure Library (ITIL). Where standard processes are available and applicable, there's no need to reinvent them using BPR or Lean with Six Sigma.

No matter where they came from, standardized processes take the form of a documented procedure that defines a series of tasks, and assigns accountability for each task to a group. It defines teamwork for a specific product or service.

Standardized processes can be very effective in the right situations.

But it's important to recognize the **limits of process engineering as a coordinating mechanism** to know where it's applicable:

- **Limited number of processes:** It takes a lot of effort to design a standardized process in sufficient detail to coordinate every team-member's work. Therefore, process engineering is typically applied to just a few major business processes.

- **Inflexibility:** Standardized processes create a rigid, assembly-line organization. The people involved and the sequence of tasks are fixed. It's only applicable to routine, repetitive processes.

- **Lack of accountability for end results:** In a standardized process, groups are accountable for their tasks; but no one is accountable for results — both those delivered to customers, and the component products and services at each step of the process. When things go wrong, it's unclear whether the fault belongs to team members or the process.

    To compensate for this, some organizations appoint a "process owner," the problems of which were documented in the case study in Chapter 3.

- **Undermines entrepreneurship:** Standardized processes define what people *do,* not what they sell. They focus people on tasks, not running lines of business that sell many different things to many different customers.

    This task-focus discourages innovation in how work gets done — there are no internal entrepreneurs who are continually looking for a better way. The lack of entrepreneurship diminishes customer focus, business planning, and innovation. It's also disempowering.

Depending solely on standardized processes results in an organization that's efficient at one or a few of its routine workflows, but perhaps less effective at the remainder of its product line. It may trim the costs of commodity products, while it fails at innovation.

Business Process Reengineering is like corporate anorexia — it makes them thinner, not healthier.

Gary Hamel and C.K. Prahalad [83]

Standardization of processes does have its place.

It's good for fine-tuning workflows within a group. In that context, it's more like a method the group uses than a mechanism of cross-boundary teamwork.

As a mechanism of teamwork, standardization of processes works best when **the organization's products and services are sufficiently standardized such that their manufacturing process can be planned in advance, and that process is well-structured so that it can be flow-charted or proceduralized.**

It works even better when **people are involved primarily in a single process** so that following the procedure fully explains how they relate to one another.

Conversely, standardizing processes is not appropriate when work processes must be flexible and tailored to unique projects, or when individuals participate in a diversity of processes. So this is not an effective approach to teamwork for most professional organizations, where each project is somewhat different, requiring a unique mix of specialists and sequence of tasks.

Mutual Adjustment

The limits of supervision and standardization are illustrated by Dr. Ed Lindblom, professor emeritus of Yale, in an eloquent metaphor. He asks you to imagine two groups of people standing on opposite street corners. When the stop-light changes from red to green, the two groups walk directly toward each other as they cross the street. What mechanism of coordination will keep them from bumping into each other?

If they were to depend on direct supervision, then someone sitting above the intersection would direct each person's every step, guiding some people to the right and others to the left as they walk. This is, of course, absurd. One person cannot process sufficient variety to guide everyone in real time.

A standardized process might tell everyone to take one step forward, stop and look ahead, and then, based on what they see, either take another step forward or a step to the right or left. This might work adequately under normal circumstances; but what happens when traffic leaves a car partially blocking the intersection? Standardization does not have sufficient flexibility to deal with ever-changing requirements.

In reality, people send each other cues (through eye contact and body language) and adjust their own behaviors to achieve their individual goals. This illustrates Mintzberg's fifth coordinating mechanism: *mutual adjustment*. This is the most powerful form of teamwork.

Conclusion

The logic is simple:

- Specialization improves performance (Principle 2). And most products and services require a variety of different specialties. So specialization depends on teamwork.

- Direct supervision creates independent silos with a fixed set of specialists in each. It creates a management bottleneck, reduces specialization, and causes rainbows and scattered campuses.

- Most organizations depend on many diverse processes — people don't do the same things every time; this is why standardization of skills or of outputs doesn't work.

- Only a subset of most organizations' processes are routine (the same steps and participants every time). This is why standardization of processes (process engineering) only applies to a subset of their challenges.

- So that leaves mutual adjustment as the most promising mechanism for coordinating teams.

The remainder of this Part describes a practical method for coordinating cross-boundary teamwork based on the concept of mutual adjustment.

Chapter 41: The Limits of Traditional Approaches to Teamwork   299

SYNOPSIS

» Depending on a boss to form teams and enforce cooperation creates a bottleneck, and leads to silo organizations.

» Standardization of skills or outputs doesn't help when an organization depends on many diverse processes.

» Process engineering is appropriate only when people are primarily engaged in a single, highly structured, and repetitive process.

» The most powerful form of teamwork is "mutual adjustment" (to be described in the remainder of the Part).

Chapter 42:
The Teamwork Meta-process:
Concepts

How can teams form dynamically, with just the right people at just the right time, and with clear individual accountabilities, a clear but customized process, and a clear chain of command?

Process for Defining Processes

The answer is found in a **"meta-process"** — **a process for defining processes.**

This meta-process is not in the traditional project-management literature. Project-management methods come later, once a team has formed and a process has been designed.

But we don't have to invent this meta-process from scratch. In fact, examples exist all around us every day.

Consider a *really* tough project.... Imagine being the project manager for a cup of coffee.

You'd have to tell the grower how many beans to plant. You'd arrange for workers to harvest the crop, and a truck to take it to the dock. You'd send a ship to pick up the beans and dock workers to load them. On the other end, you'd arrange trains to take the beans to the plant. You'd buy gas to roast the beans, and containers to package them. You'd schedule trucks to carry them to market. You'd plan shelf-space in the market, and hire check-out clerks to sell the beans. You'd arrange for water, paper filters, electricity for a coffee-maker, and a mug....

Clearly, a cup of coffee is an impossible project to manage if you have to direct the tasks of everybody on the team.

But in reality, all you need to do is buy the coffee from your grocery store. You leave it to the grocery-store manager to manage his/her staff, and to procure products from distributors. Their distributors, in turn, manage the manufacturers and shippers. And so on, all the way back to the grower.

In a market economy, each step in the value chain serves its customers and manages only its immediate suppliers. Those suppliers manage their suppliers, and so on. The *market* handles the challenges of teamwork beautifully, even on projects as complex as pouring a cup of coffee!

Market economics is the most powerful mechanism of social coordination known to mankind.

Dr. Ed Lindblom, Professor Emeritus, Yale University

Building on Principle 7, the business-within-a-business paradigm, the market model can be applied inside organizations. Here's how it works:

1. For each project or service, clear domains identify the one and only group that sells that particular deliverable. This group is considered the "prime contractor." It's 100 percent accountable for delivery of the product or service to the customer.

2. Every group is responsible for acquiring whatever help it needs from other groups (just as it does from vendors) to deliver its products and services to its customers.

   So the first job of a prime contractor is to arrange for any needed help from "subcontractors," i.e., from peers anywhere in the organization. [84]

Again, clear domains tell the prime where to find the specialists he/she needs, without having to depend on management or haphazard mechanisms such as "whom you happen to know."

3. Subcontractors are 100 percent accountable for delivering their products and services to the prime. So they, in turn, may hire subcontractors, and so on, such that processes ultimately flow across the entire organization.

For example, brand managers (as prime contractors) sell products to external clients. They may buy product engineering, manufacturing, logistics, marketing, and sales from other groups within the company. These other entrepreneurs may in turn buy procurement services, project management support, IT, etc.

Thus, processes are expressed as a series of *internal customer-supplier relationships* within an organization. At each step, one group hires subcontractors, who sell specific deliverables to this internal customer. (This nomenclature is used whether or not money actually changes hands.) [85]

Everyone lives by selling something.
Robert Louis Stevenson

All the Requirements of High-performance Teamwork

This meta-process satisfies all the requirements of high-performance teamwork:

- **Self-forming:** As each entrepreneur acquires what he/she needs from peers, teams form laterally.

- **Just the right people:** Everybody acquires just what they need, so each team combines just the right mix of specialists.

- **At just the right time:** Team members deliver specific products and services. When they're done, they're done.

- **Clear individual accountabilities:** Since people buy products and services from one another (not just "warm bodies" to work as directed), individual accountabilities and authorities are clear.

- **Clear process:** At each level of subcontracting, the buyer is accountable for integrating the subcontractors' pieces into his/her deliverable. So everything comes together to produce the end result.

- **Clear chain of command:** A prime contractor is accountable for the end result; subcontractors serve the prime; and subs to subs serve the next level up.

Of course, a prime contractor is interested in subcontractors' ideas, since they might help him/her succeed. But ultimately, the prime makes the final decisions.

Benefits of the Teamwork Meta-process

There are many benefits of this meta-process and the high-performance teamwork it induces:

- **Greater specialization:** With effective teamwork, the organization chart can be based on the Building Blocks. There's no need to structure around processes (case study in Chapter 8), or build self-sufficient silos that sacrifice specialization.

- **Empowerment:** As per the Golden Rule, every group is empowered to deliver results in the best way it can — without detailed oversight (i.e., meddling) by the prime contractor.

  And there's no need for "process owners" who disempower others by telling them how to do their jobs, but who aren't accountable for others' results (case study in Chapter 3).

- **Cooperation:** Entrepreneurs have an incentive to team with others, because they're more competitive when they utilize the right specialists rather than doing everything themselves.

  Conversely, entrepreneurs have an incentive to help peers, since internal customers are customers nonetheless, and it's important to please them.

- **Accountabilities:** It's clear who's accountable for the entire project/service. There are no "hand-offs" which confuse clients as accountability shifts from group to group.

  It's equally clear who's accountable for each sub-component of the project. This is far more effective than declaring everyone (equivalent to no one) accountable for deliverables, and betting on vague words like "team spirit" and "partnership."

- **Collaboration:** The quality of collaboration improves. People agree on what they can expect of one another. And they respect others' skills and prerogatives, and refrain from micro-managing them or stepping on their territories.

  Teams resolve disputes internally. Finger-pointing and battles for control are minimized.

  As relationships improve, collaboration improves.

- **Flexibility:** This meta-process tailors processes to the needs of each project or service — from rapid response on simple projects, to large project teams drawing on the full-time efforts of a wide range of specialists.

  There's no need for "quick versus slow" groups (Chapter 25), or rigid, pre-defined processes (Chapter 40).

- **Project management:** Managing projects is a lot easier. Of course, a prime contractor is fully accountable for the entire project. But a prime doesn't need to direct the detailed tasks of everyone on the team — a difficult challenge for large, complex projects (and impossible for that cup of coffee).

  Instead, a prime contractor only needs to coordinate the deliverables of his/her immediate subcontractors. Those subcontractors are accountable for their own deliverables, and become (sub) project managers for their components.

  Thus, the organization is not dependent on a small band of "super project managers" who take over others' lines of business when the going gets tough (case study in Chapter 37).

- **Internal customer focus:** The meta-process clearly defines whom each group's customers are. Sergio Paiz insightfully said:

  "Of course my staff report to me. But I'm their boss, not their customer. I only succeed if they serve their real customers, whether those customers are external or other managers within my company. If I'm the manager of a restaurant, our chef had better focus on pleasing diners, not pleasing me!"

- **Quality:** Internal customer-supplier relationships improve quality. In an article about the venerable Rolls-Royce Motor Cars Ltd., the New York Times said the following: [86]

  "...the company's operations were broken up into 16 zones, each comprising about 10 "teams" of employees.... The teams all have considerable leeway in how to do their jobs, and are encouraged to consider themselves customers of whatever team comes before them in the production process, thereby creating additional demand for quality. Executives here said that the system's biggest accomplishment — and the biggest factor behind the reduction in production times — was to reach Rolls-Royce's exacting specifications for quality more quickly by giving workers more freedom and responsibility to deal with production problems."

  Rolls-Royce tapped the advantages of internal customer-supplier relationships and empowerment. The resulting quality of their products is world renown.

- **Leadership:** Since executives don't have to personally form and manage project teams, they can focus on more strategic challenges such as ensuring that the organization is functioning well and heading in the right direction.

Chapter 42: The Teamwork Meta-process: Concepts

This business-within-a-business approach to teamwork builds explicit, but flexible, cross-boundary processes. It defines clear individual accountabilities. It builds close partnerships based on mutual interdependence. And it supports an organization of empowered entrepreneurs, in keeping with all the Principles.

SYNOPSIS

» A "meta-process" is a process for defining processes that are tailored to the needs of each project or service.

» For each project or service, a "prime contractor" buys help from any needed subcontractors, who in turn may buy help from others.

» This meta-process, based on internal customer-supplier relationships, accomplishes all the goals of high-performance teamwork.

Chapter 43:
The Teamwork Meta-process:
Mechanics (Walk-throughs)

Implementing the teamwork meta-process involves three components: preparations (one-time); a pragmatic procedure for initiating projects and services (ongoing); and a supportive culture.

Preparations: Product/Service Catalogs

The teamwork meta-process is built on the notion of a prime contractor "buying" from subcontractors. What do people buy from one another? *Products and services.*

Thus, each group should define its *catalog* of products and services — what it sells to clients and as well as to peers within the organization. Each item in the catalog is *something a customer will own or consume,* not the tasks that go into making it.

The meta-process can be done without catalogs. But catalogs help by providing a clear, precise language for defining subcontracts.

Beyond that, catalogs have many benefits, including the following:

- They're a language for clarifying commitments to customers.
- They define everyone's accountabilities for results, not just for tasks or competencies.
- They reinforce customer-focus by causing staff to think about what their customers will get from them.
- They encourage empowerment. People can be managed by

their products and services (results), without attempting to control the tasks they perform to produce those results.

- They bring domains to life, and induce an entrepreneurial spirit (e.g., "What else might I sell?").

- Catalogs are the first step in implementing internal market economics, where products and services have a cost and clients' expectations must fit within available budgets. [87]

Pragmatic Procedure: Walk-throughs

What, exactly, do you want staff to do when a new project is initiated? The very first step should be a "walk-through."

A walk-through is **the hierarchy of deliverables for a specific project or service** — prime contractor to subcontractors, subs to sub-subs, and so forth.

It defines who's on the project team, the specific results each is accountable for, and the chain of command within the team. It's like a high-level project plan. But it's distinctive in that roles are defined in terms of *deliverables* (products and services), not tasks.

Figure 26: Format of a Walk-through

Customer	Product/Service	Supplier
Client	Solution	Prime
Prime	Subcomponent	Sub 1
Sub 1	Service	Sub 2
Prime	Service	Sub 3

Start with a chart (perhaps a spreadsheet) with three columns, as in Figure 26. (You could add a fourth column for notes.)

Here are the steps in doing a walk-through:

1. Define exactly what the end-customer wishes to buy. This is noted as the first row on the chart:

    Column 1: Customer — for the prime contract, this is the end-customer, perhaps a client.

    Column 2: Product/Service — the specific item the end-customer will get, drawn from the prime contractor's catalog.

    Column 3: Supplier — the prime contractor group. It should be easy to identify the prime contractor based on clear domains (the organization chart).

2. The prime contractor decides what to buy from subcontractors. Each subcontract is represented as another row on the chart, with a customer (now the prime), the name of the internal product or service (from the subcontractor's catalog), and the name of the supplier (the subcontractor).

3. Each subcontractor then gets a chance to buy any sub-subcontracts it needs. Each is represented as another row on the chart.

    This goes down as many levels as are necessary.

    Note: Only enter direct subcontracts, not internal support services that you're already buying for other reasons. Those indirect services should be treated as separate walk-throughs.

A hierarchical project plan emerges. Each row on the chart is an internal "contract" between a customer and a supplier.

Example of a Walk-through

Here's a simplified example of a walk-through. (Remember, the word "buy" is figurative, whether or not money changes hands.)

In a consumer products company, Brand Managers are accountable for the profit/loss of product lines.

- Brand Managers buy market research from Marketing to assess the opportunities for new products, and define customers' requirements.
- Brand Managers buy product designs from Design Engineering.
    - Design Engineers buy help from Manufacturing Engineers to ensure that designs are feasible to produce.
    - Design Engineers buy help from Materials Engineers (or other Design Engineers for component parts).
    - Design Engineers buy support from the Project Management Office.
- Brand Managers buy raw materials from vendors, with the help of Procurement.
- Brand Managers buy production services from Manufacturing.
    - Manufacturing buys plant set-up from Manufacturing Engineering.
        - Manufacturing Engineers buy support from the Project Management Office.
    - Manufacturing buys transportation for raw materials and finished goods from Logistics.
- Brand Managers buy storage of their raw-materials and finished-goods inventories, and shipment to customers, from Logistics.
- Brand Managers buy marketing communications from Marketing.
    - Marketing buys graphic arts services from Administration.
- Brand Managers buy sales-force time from Sales.

And, of course, everybody buys support services such as IT, HR, Finance, Facilities, Legal, and Administration. But those are separate walk-throughs.

When to Do Walk-throughs

Your first few walk-throughs might map current (or completed) projects, so that managers can practice on known examples.

Once managers understand the process, walk-throughs should be mandatory at the kick-off of every new project or service.

You might also do walk-throughs for existing services to clarify and optimize processes.

At first, walk-throughs may feel like learning a foreign language. But after a bit of practice, the process becomes intuitive and teamwork across organizational boundaries seems natural.

Initially walk-throughs are done in management team meetings.

As managers gain experience, walk-through meetings may require only the relevant groups. Ultimately, walk-throughs may be done asynchronously, with a series of one-to-one conversations between internal customers and suppliers.

The... process forced us to specify our products and services,
[and] our customers and suppliers
for each operating and support group within our [department].
This brought to light many of the inefficiencies that we had been
struggling with and allowed us to deal with them.
William T. Houghton, President, Chevron Information Technology Company

Culture Conducive to Teamwork

The third component of the teamwork meta-process is culture.

By culture, I don't mean people's values, attitudes, feelings, or beliefs. A more effective and pragmatic way to express, and influence, culture is through behaviors. [88] Unlike values, behavioral principles can be taught, practiced, and measured.

Here are just a few examples of cultural principles that are conducive to teamwork:

- We do not sell products or services outside our domain.

- We never make a commitment we can't keep; and we fulfill every commitment.

- We never make a commitment for others; we arrange needed subcontracts before we make a commitment to our customer.

- We subcontract for deliverables, not just for people's time (unless all that's needed is a bit of advice).

- We treat commitments to internal customers as just as obligatory as commitments to clients.

These are just examples. A comprehensive culture defines expected behaviors in many themes, including ethics, integrity, entrepreneurship, customer focus, empowerment, cooperation, as well as teamwork.

An explicit, well-defined culture can reinforce the new way of working in a new structure.

(More examples of behavioral principles that reinforce a principle-based structure are available in the Supplement to this book; contact <ndma@ndma.com>.)

SYNOPSIS

» Catalogs provide a precise language to define internal customer-supplier relationships.

» The teamwork meta-process is implemented through "walk-throughs". A "prime contractor" forms a team by subcontracting for needed help.

» Subcontractors are accountable for their deliverables, and may acquire help from others as needed. Thus, teamwork spreads laterally through the organization.

» Walk-throughs should be the first step in initiating every new project or service.

» Walk-throughs can also clarify processes for existing services.

» Culture can reinforce teamwork. Effective culture is expressed as clear principles of behavior (not values).

~ PART 9 ~

The Process of Restructuring

> *What we have to learn to do,*
> *we learn by doing.*
>
> Aristotle

To this point, you have the Principles, Building Blocks, diagnostic method (the Rainbow Analysis), design method, and the teamwork meta-process. But there's more to a successful reorganization than a carefully designed organization chart. An effective implementation process is just as important as having the right answer.

How are restructurings usually done?

The boss draws one or two tiers of boxes, typically with the capabilities of the current senior managers in mind.

Or the senior leadership team discusses the boxes until they agree, although it may not be clear what they're agreeing to, since they all understand the few words in the boxes somewhat differently.

Then, lower-level groups are assigned to the new top-level boxes, for the most part, moving existing groups intact.

Finally, the organization chart is announced, perhaps with some presentations that are intended to explain the intent of the new structure. And with that, senior managers declare victory and the staff are left to figure out how to make it work.

The combination of limited involvement and unclear language means that leaders, staff, and clients have vague and varying perceptions of what was intended. But when people don't understand the logic behind the structure, they revert to old habits and nothing much changes.

Furthermore, it may assume teamwork; but just announcing an organization chart does nothing to improve cross-boundary teamwork (no walk-throughs). With the same old constraints on teamwork, staff do whatever it takes to get their jobs done, and silos evolve despite what the new organization chart says.

After months sorting out how things should work, during which confusion is embarrassingly visible to both staff and clients, little changes. People fall back into old patterns of work, until a year or two later, when the organization repeats the entire process.

There is a better way....

Chapter 44 describes the basic underpinnings of an effective change process.

Chapter 45 describes a proven reorganization project plan.

Chapter 46 addresses a common stumbling-block for internal service providers: establishing an internal Sales function.

Chapter 47 looks at mergers and consolidations, where two (or more) organizations are integrated into one. The science of structure provides a powerful, fact-based way to integrate the structures, and to glean the best from both organizations.

Chapter 44: Change Management

When planning a restructuring, some things are obvious. Careful thought has to go into the design of the organization chart. Of course, the selection of people for those new roles is critical. And there are myriad administrative details that must be handled.

Other things, while less apparent, are equally critical. Staff have to understand how things are supposed to work, learn to think like entrepreneurs, and institutionalize the teamwork meta-process.

The implementation process must accomplish all this, and more. It should be a positive experience that builds support — hopefully even enthusiasm — for a new way of working. It should *capture hearts and minds*.

The key to capturing hearts and minds, and the topic of this Chapter, is effective change management. [89]

Chapter 45 then puts all the pieces together in a comprehensive restructuring project plan.

Challenges of Change

People often feel threatened by a restructuring. They fear they may no longer be competent in a new job. They may feel they're losing control of their destiny, or even at risk of losing their jobs.

If staff are too uncomfortable with the change, they're likely to resist it. But the new structure depends on people embracing it and adopting new ways of working.

Beyond the initial acceptance, the change has to be sustained until it's "forgotten" in the sense that it's just the way we work around here, and is no longer a "change."

Change management is not a parallel process. It's built right into the design of the restructuring process. There are five basic elements of change management: climate for change, participation, "ground-rules," transitions, and communications.

Climate for Change

There are three components that create a "climate" conducive to change:

- **A clear mandate for change** — the so-called "burning platform": Even if the future might be better, people aren't going to embrace change (with its costs of disruption and risks) unless the status quo is unacceptable.

    The Rainbow Analysis (Chapter 17) can help staff see the need for change.

- **A vision of the destination:** People may understand that change is required; but they're not going to "jump from a hot pan into the fire." They need to know where the change will take them, a better future that makes change worthwhile.

    The Principles and Building Blocks can create a vision of the end-state.

- **A clear path forward:** People need to believe that the change will work before they support it.

    By seeing a well-planned implementation process, staff will feel more confident that the restructuring will succeed.

Participation

Participation in the change process is invaluable. As Peter Senge said, "People don't resist change. They resist *being* changed." [90]

A participative process involves more than just the top few executives. It engages the entire management team, and ultimately, all the staff. Like a barn raising, the community works together to design and implement the kind of organization in which people want to work.

Participation is important for at least four reasons:

1. It brings into the design process detailed knowledge of "the way things really work around here." The diversity of the management team's experiences, perspectives, and insights adds a richness to the process, and ensures that the resulting design is comprehensive and feasible.

2. Participation builds understanding. A structure designed around businesses within a business may be radically different from the past. Without a deep understanding of this philosophy, managers won't see their jobs as significantly different; and the new structure won't work as planned.

3. By seeing all the various lines of business within the organization, people become aware of the help they can get from peers, countering any historic need for self-sufficient silos.

4. Participation engenders commitment. It builds a sense of ownership and responsibility for making the new structure work.

Of course, the image of your entire management team — dozens, if not hundreds, of people — debating the boxes on an organization chart is absurd. Widespread participation must be carefully orchestrated so that discussions don't descend into a battle of opinions clouded by personal interests, territorialism, and self-serving politics.

A well-planned process takes the design team through a series of decisions, one at a time, each building on the last.

At each step, the Principles and the Building Blocks induce healthy debate, shifting discussions away from politics and parochial self-interests to objective analysis and fact-based decision making.

In this process (described in Chapter 45), teams of 20 to 30 leaders can come to consensus on a new organizational structure in just a few days. Then, a much larger group of managers can expand on that high-level design, and participate in the walk-throughs that bring the organization chart to life.

"You teach me, I forget.
You show me, I remember.
You involve me, I understand."

Edward O. Wilson, sociobiologist

Ground-rules

The very nature of the change has to be positive, not threatening. Staff should feel enthusiastic, not fearful.

I've found that three guarantees (ground-rules) are necessary to create a safe environment that's conducive to open participation, and to elicit everyone's commitment to the new organization:

1. **No one will lose employment** because of the restructuring. Any necessary headcount reductions should occur before (or well after) structural change. (Of course, individuals can still be fired for other reasons.)

2. **No one will lose compensation** as a result of restructuring. If people fit best in lower-graded jobs, their compensation will be protected, at least for one or two years to give them time either to grow the job into something that warrants their compensation or to find another job.

3. **No one will be forced to relocate** (geographically) as a result of restructuring. If people fit best under a manager who lives elsewhere, they'll be managed remotely.

Transitions

Bill Bridges coined the term "transitions" to mean the feelings people experience when change happens around them. [91]

Bridges identified three phases: First, we mourn the loss of the old, familiar ways ("endings"). Once we let go, we feel disoriented and don't know how to behave ("the neutral zone"). Finally, as we figure out how to succeed, we take charge of the new order ("beginnings").

It's essential that people don't get stuck in the endings or neutral-zone phases. Beyond that, there's much leaders can do to help staff (and themselves) deal with each phase constructively. Training in transitions management can be built into the change process.

Communications

Open communications throughout the process is essential. It quells fears, builds trust, engenders support, and develops the perspectives and knowledge that are critical to successful implementation.

In general, **the more open the communications, the greater the likelihood of success.**

There are limits, however. Leaders must **never engage in speculation.** Doing so only feeds the rumor mill and can set inappropriate expectations, if not raise unfounded fears. If a question has not yet been decided, the right answer is, "We don't know yet."

At every step in the restructuring process, the leadership team should craft its communication plan, sharing openly all it knows while avoiding speculation.

The style of communications is also important. The language should be straightforward, compelling, and positive about the future. [92]

Reinforcement

Throughout the process and beyond, the new paradigm must be reinforced to make the change stick.

In a sad example of the contrary, a CIO in a university was very successful at implementing a new structure and the new way of working that went with it. But he didn't reinforce it.

He didn't clearly describe the business-within-a-business paradigm to candidates for some open leadership positions, and didn't make support of that new way of working a key selection criterion.

And when some senior leaders that he hired circumvented the new structure, he didn't treat it as a serious performance problem.

When he retired, one of those new-hire executives was appointed "acting" leader (while the institution searched for his replacement). Before she was replaced, she unilaterally returned the organization to its prior structure — over the objections of those senior leaders who had been part of the process from the beginning, but with the support of some leaders hired after the change process.

To make the change stick, you must reinforce it in every possible manner — in your hiring criteria, new-hire training, performance appraisals, and firm performance management.

Reward systems (e.g., performance appraisals, bonuses) should reinforce customer focus, entrepreneurship, and performance. For example, 360-degree reviews can provide input to performance reviews from customers and suppliers, as well as subordinates. [93]

Reliable delivery can be measured in the context of "contracts" for products and services. And rates can be compared to competitive benchmarks.

A well-designed structure simplifies the design of reward systems. It aligns individuals' best interests with the enterprise's best interests. So there's no need to design incentives for altruism.

And there's no need to tie individual incentives to team performance (which they can't completely control). Staff already have incentives to serve internal customers on teams; that's part of their own success and individual metrics. So they're already rewarded for team success, thanks to walk-throughs.

Thus, rewards can be tied directly to things people can control, especially serving their customers well.

Making the change stick requires a strong commitment from the top executive, at least for a few years until the new way of working is "just the way we work around here."

SYNOPSIS

» Effective change management is not "added on" to a change initiative; it's inherent in the design of the implementation process.

» Change management creates a climate for change: dissatisfaction with the status quo, a vision of the destination, and a clear path to get there.

» Change management includes effective participation, with "groundrules" that help people feel safe.

» Change management can directly address transitions (the feelings people experience when change happens).

» Change management includes open communications.

» The change must be reinforced as a performance expectation.

Chapter 45:
The Restructuring Process, Step by Step

This Chapter puts all the pieces together — the Principles, Building Blocks, diagnosis and design methods, teamwork meta-process, and the challenges of change — into a step-by-step implementation project plan.

Pay Me Now, or Pay Me Later

This process involves a lot of planning before the new structure is deployed. In my experience, it's more effective to get everyone to understand not just a new organization chart, but also a new way of doing business, *before* it's deployed, not afterwards.

Sure, this takes time. But it's "pay me now, or pay me later." A true story illustrates the importance of patience:

Noel Thomas, former strategic planner at a very large North American steel company, visited a Japanese steel mill to learn how they consistently delivered superior products at lower costs. He expected to find exotic new technologies. Instead, he was surprised to find the same kind of mills and production processes as his company used.

After some study, he came to the following insight: In Japan, they spend (figuratively) three months planning, two months building, and one month commissioning a new production line. North American companies typically spend one month planning, two months building, and then <u>years</u> trying to get a new facility running well.

Noel's insight applies to many things in life, including an organizational restructuring. A participative planning process requires significant time and effort up front. But it's a good investment. Thanks to the time spent planning and educating leaders before the official change-over, the new organization "hits the ground running."

And if you consider the point in time when the new structure is really operating as it should, this well-planned, participative approach is ultimately quicker. It's another example of the old fable where the tortoise beats the hare.

"We don't have the time to do it quickly!"
Dave Anderson, then CIO, later President, American Family Insurance

---

Overview of the Implementation Process

A proven and extremely well-documented implementation process has evolved through the experiences of dozens of diverse organizations over more than three decades. [94] It delivers both an organization chart and the teamwork meta-process, and it incorporates all the elements of effective change management.

I'll use the case study described in the Foreword (by Sergio Paiz of PDC) as an example as we go through the process, step by step.

In large organizations, the process first develops the tier-one structure (reporting directly to the top executive). Then, selected tier-one leaders cascade the process to the next tiers of leadership. Sergio, with his 850-person organization, took this approach.

Chapter 45: The Restructuring Process, Step by Step 327

In small organizations, the process can be condensed by designing the top two or three tiers all at once.

Figure 27 is an overview of the process.

Figure 27: Overview of the Restructuring Process

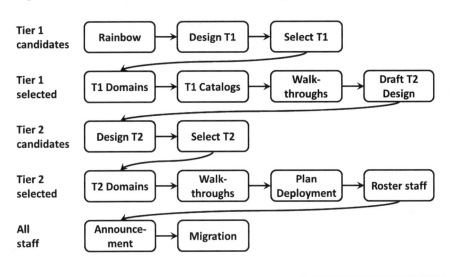

There are two things to keep in mind as we discuss the steps in the restructuring process:

One, at every step of this process, a communications plan is crafted.

Two, while this planning is occurring, the old organization and leadership team remain in place and continue to run the business.

Tier-one Design

The process starts with all candidates for tier-one leadership jobs (those who hope to report directly to the top executive). All incumbent tier-one leaders are included, but the team isn't usually limited to them. Eligibility for participation is defined by a set of rules (not favoritism), or through an application process.

In Sergio's case, this amounted to 13 leaders (as well as himself and a few corporate observers).

In smaller organizations, all candidates for the top two or three leadership tiers are included.

That "design team" studies the science of structure, and applies the Rainbow Analysis (Chapter 17) to their current organization chart.

In this first workshop, Sergio's leadership team graphically saw how every Building Block was scattered among its divisions and corporate staff. It was a very colorful organization chart! This cemented the need for change.

Then, participants individually draft proposed organization charts. There's no requirement to structure around the Building Blocks; but boxes have to be labeled with the Building Blocks that they contain so as to communicate clearly their intent.

In the next workshop, the team is facilitated to consensus on the tier-one structure. This is not a matter of voting for your favorite proposal. First, the detailed lines of business that must exist somewhere in the new structure are extracted from all the proposals. Then, the way each chart clusters those lines of business is examined. Resulting clusters represent tier-one jobs, defined by the list of lines of business under each.

Chapter 45: The Restructuring Process, Step by Step

Parts 6 and 7 provide detailed guidelines for this design step.

This is when Sergio told his leaders to drop their business cards at the door and join him in designing a new organization chart, starting with a blank sheet of paper. The symbolism of this reinforced that the leaders were expected to let go of their current fiefdoms, and think about what's best for the whole company.

Sergio's leadership team decided by consensus to consolidate all internal lines of business to maximize specialization. They defined 11 tier-one jobs that roughly followed the Building Blocks.

"Once we adopted the... Principles, we had a framework that provided a basis for any decision, defused the inevitable turf battles before they ever began, and accelerated our learning process in a way that probably could not have occurred otherwise."

John Benci, General Director, MIS, The Canadian Wheat Board

Tier-one Selections

With the boxes and no names, the next step is a fair selection process in which the top executive chooses people for those tier-one jobs. Choosing from the pool of people who designed the new structure ensures that selected leaders have a deep understanding of, and commitment to, the new organization. [95]

Generally, the executive asks each candidate for his/her job preferences, and considers that when making selection decisions. Having been part of the design, and then having some say in one's own destiny, builds tremendous commitment to making the new structure work.

Sergio did this, meeting with everybody on the design team. Most leaders opted for something similar to what they'd been doing. But a few surprised him and asked for a fresh start in a new area.

Sergio filled 6 of the 11 tier-one positions with existing leaders. But 5 of the boxes remained open because he didn't find the needed talent internally. He initiated external searches for the empty positions.

This is also a good timeframe to provide training that helps leaders deal with their own transitions.

Tier-one Domains

Before the new structure is deployed, this newly appointed tier-one leadership team (at this point, still just a planning team, not official yet) plans the details of how the organization will work.

First, they document their domains. Domains are a precise definition of groups' boundaries, phrased as constraints on what

Chapter 45: The Restructuring Process, Step by Step   331

each group sells. With clear domains, you can look at any project or service and know which is the one and only group that sells that particular thing.

Beyond just producing today's products and services, each group is accountable for the long-term viability of its business within a business. Domains are written in a way that describes their business, now and in the future.

Domains are not only an essential part of the definition of the structure (Principle 3). The process of writing them helps newly appointed leaders understand, and develop ownership of, their new jobs.

Next, leaders may translate domains into their catalogs of products and services.

Like most, Sergio's team struggled at first, tending to define boundaries and catalogs by what they *do* rather than what they sell. But after a series of coaching workshops, most came to see their jobs as businesses within a business, chartered to produce products and services for others.

Tier-one Walk-throughs

The second essential output of an effective restructuring process is installing the meta-process for forming teams and defining processes — walk-throughs (Part 8).

Leaders practice walk-throughs until teamwork becomes second nature. Walk-throughs help them "debug" the structure, clarify domains, and understand how work will actually get done.

Cascade to Next Tiers

In large organizations, the next step is to "cascade" the process to the next tier.

With a solid understanding of how the new organization will work, the future tier-one leaders draft the tier-two (and sometimes tier-three) organization chart.

Then, candidates for these next-tier leadership jobs are engaged in the process. This may be a large group, but their input is carefully orchestrated in a workshop in which tier-two candidates learn the science, see the need for change through the Rainbow Analysis, and understand and refine the tier-two structure proposed by the tier-one team.

In Sergio's case, the candidates for tier-two leadership positions amounted to approximately 60 people. He and the newly selected tier-one leaders invested time in educating them in the process to date, and gathered their input on the draft tier-two and tier-three organization chart.

Next-tier Selections

Once the remaining tiers of structure are finalized, a fair selection process assigns people to those jobs.

Tier-one leaders make the selections *as a team*. Thanks to the walk-throughs, they understand their interdependencies. So instead of vying for the best people, they work as a team to place people in the jobs where they'll contribute the most. Again, candidates' preferences are taken into account.

Next-tier Domains and Walk-throughs

Once selected, the new tier-two leaders document their domains and catalogs (dividing up their bosses' tier-one domains and catalogs). Then, they practice walk-throughs until they, too, understand exactly how real work will get done in the new structure.

As the selected leaders in Sergio's organization practiced walk-throughs, they got a much clearer sense of their individual accountabilities as well as their interdependencies. And they planned the details of how actual processes will work in their new structure.

When you get to the point where any tier-one or tier-two leader can be presented with a real-life project or service, and is able to describe a walk-through in the same way that every other leader would, then finally the new leadership team is ready for roll-out.

Roll-out

When they've done enough walk-throughs, leaders plan the process by which the new structure will be deployed. There are myriad details to think about.

Also during this timeframe, leaders may be trained to help their staff deal with their transitions.

Toward the end of this step, they roster the remaining staff, assigning people based on what they currently do (not their preferences, since the same work has to continue to be done by those same people).

When everything is ready, the new structure is made official. This generally takes the form of all-staff meetings, so that everybody has a chance to understand the new organization, including what drove the change, the principles behind its design, the details of the new organization chart, walk-throughs, the benefits to them, and the next steps in the process (migration).

Sergio's leadership team did all the necessary preparations. Then, on the announcement day, they presented the new structure to the staff in an all-staff meeting, with video teleconferencing to remote sites so that all employees could participate.

Migration

Finally, all staff are engaged in understanding the new structure and migrating into it. The learning and migration processes are carefully planned to ensure smooth and lasting change, with a minimum of disruption and risk.

Everyone's commitments are documented; and if they no longer fit one's domain, they're carefully transitioned to the appropriate group.

At PDC, as one would expect, the reaction was mixed. Sergio said, "I guess people thought I was crazy. People get accustomed to their previous experiences, and it was difficult for most to envision a better way to do things."

Sergio spent a lot of time with his leaders. Many thrived in the new environment. But some found it difficult to accept the change and left the company. As Sergio recruited new talent, he was careful to make sure all new hires were supportive of the structure and the business-within-a-business paradigm.

After a few years of reinforcement and hiring entrepreneurial talent, with minor adjustments to the structure (always considering

Chapter 45: The Restructuring Process, Step by Step   335

the Principles), Sergio's company is fully up and running in the new paradigm. As he reports, "Over time, people understood it more. And now people are convinced this is absolutely the way."

(For a detailed project plan, qualified executives may contact <ndma@ndma.com>.)

SYNOPSIS

» A participative implementation process maximizes understanding and commitment, and careful planning before deployment ensures that the new structure "hits the ground running."

» The restructuring process begins with tier-one jobs, and then cascades downward through the layers of structure.

» At each tier, candidates participate in the design, and express their preferences as to the jobs they'd like.

» Once selected, leaders craft clear domains and catalogs, and then practice walk-throughs. The organization doesn't go into effect until the leadership team fully understands how it will work.

» The migration process ensures that all accountabilities are moved to the right place, without missing any commitments and with a minimum of disruption.

Chapter 46:
Establishing the Sales Function in an Internal Service Provider

For internal service providers, the Sales function (or whatever politically correct name it may adopt) is critically important. But historically, in many organizations, it's been either neglected or considered a part-time job for senior leaders.

Hopefully, at this point, you're convinced that Sales is a profession in its own right. If so, you may have established it as a new group within the organization. That's wise.

But many fail. A new Sales group is often resisted by both clients and peers within the organization. After a short period of time, many are disbanded.

It's important to understand why they fail. You wouldn't want to give up on an important concept due to poor execution. And there's no need to repeat the mistakes of the past.

This Chapter looks at the hurdles, and suggests how to successfully set up a new internal Sales function.

Twelve Reasons Internal Sales Functions Fail

Here are the twelve most common reasons why internal Sales functions fail:

1. The function is staffed with the **wrong people** who lack sufficient stature or business knowledge to command the respect of senior client executives, and to engage in intelligent business discussions at that level.

2. Sales staff **fail to add value** for lack of training in methods such as opportunity discovery, consortium facilitation, and benefits estimation. If they're nothing more than a friendly point of contact, the group may not be worth its cost.

3. They become a **communications bottleneck** by demanding that all client communications go through them. Sales should be a default point of contact, not a single point of contact.

4. They **make commitments for others.** Others may not be able to deliver, and invariably resent Sales staff for managing their time and setting them up to fail.

5. They design solutions (perhaps at a high level, but designs nonetheless), **encroaching on the prerogative of Engineers.**

6. They make decisions on behalf of clients or speak for clients, **encroaching on the prerogative of clients.** This may lead to projects in which Sales is the sponsor rather than a client, and the lack of real client sponsorship dooms projects.

7. They see their role as **defending clients' interests "against" the rest of the organization,** rather than facilitating collaboration between the two communities. This leads to an adversarial relationship with peers. Of course, without the cooperation of others, Sales staff can't get anything done.

8. Others within the organization **do not see the value** of the function, and may even resent Sales staff for "taking away" an interesting part of the jobs they used to do. To gain peers' support, everyone must understand how their own success depends on the success of the Sales function.

9. **Engineers remain substructured by client** and refuse to relinquish their role as primary client liaisons. (This problem

is rooted in an inappropriate substructure in the Engineering function. It generally only occurs when a Sales group is added to an existing structure, without a Clean Sheet design.)

10. They **charge for Sales services that should be free.** Clients choose not to pay because they haven't yet learned to appreciate its value, and the function is eliminated for lack of funding. (By the way, vendors don't charge for sales time. Most Sales services should be considered "overhead" and built into everybody else's rates.)

11. Engineers have not reserved **sufficient "unbillable" time** (i.e., non-project time) to develop proposals for new projects. Sales staff may get clients excited about high-payoff opportunities, and then find that Engineers are too busy to propose alternatives and estimate costs.

12. The priority-setting process makes it **difficult to get small, well-focused strategic projects funded.** If the Engineers are tied up with long-term projects, Sales cannot add value by identifying small, high-impact projects. In such situations, there's little for Sales to do other than relationship building.

Four Keys to Success

So what does it take to set up an effective internal Sales function?

There are four keys to success: the right people, training, a business-startup strategy, and cooperation with peers.

People: First and foremost, it takes the right people.

Sales staff have a foot in both the organization's world and the clients' world. They're business people first — comfortable talking with key client executives about their business strategies, not just about the organization's products and services. They're the sort of up-and-coming executives that clients would love to steal for management positions in their business units.

To gain access to influential client executives and add value to their business thinking, Account Sales staff must be seasoned and well connected. They have the political savvy to determine which individuals are most critical to clients' strategies, and hence most worth serving. These are not "business analysts." They're senior leaders, often at the same grade-level as other tier-one leaders.

Sales staff understand the organization's products and services as a "smart buyer," but they're not experts in any subset of the product line (to avoid being biased). They're not accountable for the design or delivery of others' products and services.

Especially important, Sales staff are specialists in the linkage between clients' businesses and the organization's products and services, including knowing how other companies use its deliverables to leverage their strategies.

Training: Next, it takes training in the role and its methods.

First, Sales staff must understand their own catalog of services. By defining their services, Sales staff come to understand what they do, and (equally important) what they don't do. This helps them explain their value to clients and peers.

Many of their services require training in methods. The biggest payoff is typically found in the method by which they analyze clients' business strategies and help clients discover high-payoff opportunities for the organization's products and services.

In addition, Sales staff need excellent interpersonal skills, communications skills, and emotional intelligence. They need behavioral skills such as listening, interviewing, meeting facilitation, conflict resolution, and counselor selling. And they're skilled in negotiation, and in driving win-win deals to closure.

And of course they must understand the organization's structure and be aware of everybody's products and services.

Startup Strategy: A new internal Sales function must be deliberate about its start-up strategy.

Reviewing its own catalog, a new Sales group should decide which services they'll deploy first, second, and so on. It's important to quickly add value by focusing on the one service that's most needed and can be delivered with a modicum of training.

Furthermore, a start-up strategy should focus on clients who are most likely to get value from them — people of sufficient level to impact key business goals, and who are open to discussing creative uses of the organization's products and services.

Cooperation: Of course, cooperation between Sales and the rest of the organization is essential. Peers must understand why it's in their own best interests for the Sales function to succeed, appreciating what's in it for them. An internal "marketing" (education) campaign may be needed.

One benefit to peers is improved client relationships, and resolving any interpersonal or political issues. Sales relieves them of these distractions so that they can focus on their work, which they no doubt find more rewarding.

Sales also helps everybody succeed by clarifying contracts with clients, and by bringing them strategic opportunities to work on.

In addition to building awareness of its value, the Sales function should establish clear boundaries between themselves and peers.

A key boundary is between the Sales "opportunity identification" and "requirements analysis" services and Engineers' solution-design work. Sales should define the "what," but leave it to Engineers to determine the "how." The level of detail of their definition of the "what" is the boundary which should be agreed.

A good way to get a new Sales function started is a workshop in which the team defines its catalog of services, analyzes the skills and methods it needs, decides its start-up strategy, and develops a communication plan for both clients and peers.

A strong next step is a workshop with peers, especially Engineers, to reach a consensus on their boundaries.

Then, Sales has to quickly get out and deliver real value to clients and the organization.

SYNOPSIS

» For internal service providers, Sales may be a new function; it must be established carefully.

» There are twelve reasons many Sales functions fail; awareness of them helps you avoid the mistakes of the past.

» The four keys to success are: the right people, proper training in the methods of Sales, a deliberate start-up strategy, and soliciting the cooperation of the rest of the organization.

Chapter 47:
Consolidation of Shared Services and Acquisition Integration

When companies merge or acquire one another, and when decentralized functions are consolidated into shared-services departments, two (or more) organizations are melded into one.

A consolidated organization is supposed to perform better than two smaller organizations. There should be savings through economies of scale and the elimination of redundancies, better performance through more specialization, and synergies.

Unfortunately, consolidations don't always produce the expected benefits. Just jamming the pieces of one organization into another and firing some people (sadly a common approach) may amount to nothing more than cutting costs by reducing the quality of services and increasing risks, while gaining few of the promised synergies.

This Chapter explores what "success" means in an organizational integration process, and how to attain it.

Benchmarks of a Successful Integration Process

First, let's reflect on what success looks like. I believe that in an effective organizational integration process, we do the following:

Commitments

- Continue to meet all existing commitments (until they are explicitly changed).

- Make new commitments with accommodation for the time to be invested in the integration process.

- Make deliberate decisions about which deliverables (if any) are to be cut; don't just cut costs and let things fail randomly.

Structure Design

- Base decisions on principles and facts to get the best of both entities, not simply winners dominating losers.

- Design the new organization to meet customers' expectations of results, rather than their opinions about its structure (e.g., demands for groups dedicated to them).

Treatment of People

- Engage leaders from both groups in the integration process.

- Treat everybody equitably. Assign jobs based on competence, putting the best talent in each job regardless of its origins.

- Whenever possible, manage global groups rather than forcing people to relocate.

Change Management

- Openly communicate with everyone concerned, including: the vision for the new organization, the scope of who's affected, the integration process, its timetable, its status, and how the process will affect people individually.

- Help people deal with the personal stresses of change (their "transitions"). [96]

Taking Advantage of Change

Beyond all those basic expectations, change opens up broader possibilities. While you're changing, you might as well do it right.

There's an old saying that a newly wed couple shouldn't move into his house or hers, but rather should build a new home together. You can use the integration process to build a new, high-performance organization, beginning with a clean sheet of paper.

This may be more than an opportunity; it may be a necessity. Just combining two structures may result in an organization that performs no better (or even worse) than either of its components. Meanwhile, expectations may be higher, given the promise of cost savings and synergies. "1+1=2" generally isn't good enough.

On the other hand, by applying the science of structure, the merged organization can become capable of significantly higher levels of performance than just the sum of its parts.

Patience, Synergies Take Time

Before describing an effective integration process, one caution is in order. It takes time to deliver the promised benefits.

Combining two organizations does not, in itself, save money. Even where there are redundancies, costs may not be reduced. The volume of work doesn't go down, so a single organization must deliver just as much as the sum of the two separate groups.

There may be savings in economies of scale. Examples include sharing licenses, exercising the buying power of a larger organization, and reducing "safety stocks" (any excess capacity needed to handle peak loads). But these easy savings typically fall far short of hoped-for benefits.

There may be savings in simplifying the work. For example, in IT, two companies may share a single ERP application rather than running two separate systems. But, of course, this requires adopting common business processes, so it takes time and money to implement the change and harvest the savings.

Another, perhaps more lucrative, form of synergy comes from Principle 2. A larger organization can afford a higher degree of specialization, which leads to lower costs, higher quality, more innovation, and greater agility.

And, in the case of corporate mergers and acquisitions, there may be product and market synergies that increase revenues.

But all these savings and synergies require time and money to implement. They depend on the new organization taking advantage of the strengths of each of the pieces, and performing well as a single, integrated organization.

Patience is key. **If you rush to the savings, you may never get them.**

If you cut headcount and costs before the organization is ready, a small but no more effective (indeed, temporarily less effective) organization is bound to reduce its deliverables, quality, agility, and innovation, while taking greater risks. Things will break.

Furthermore, it may prevent you from making the investments that are necessary to produce the real benefits.

Meanwhile, rushing the process creates chaos and turnover, and ultimately it will take even longer to achieve a functioning end-state.

Patience and a well-designed process should more than pay off in terms of both the ultimate benefits and the total time it takes to get there.

The C-I-O Strategy

An effective organizational integration process has three distinct phases, each with its own challenges and processes, captured in the acronym "C-I-O":

1. **Consolidate:** Put the two groups together under a common boss, leaving their prior structures intact. While this alone gains no synergies, it gives one executive legitimate authority to lead the integration process.

2. **Integrate:** This is when the structures are merged. But staff continue to do what they've been doing so as not to miss any commitments.

3. **Optimize:** Then, each manager optimizes his/her newly integrated group to produce the expected cost savings and synergies. This phase is ongoing.

Re-emphasizing the need for patience, note that the benefits aren't delivered until the Optimize phase.

Consolidation Phase

In the Consolidation phase, the two (or more) organizations are placed under a common boss. This gives that boss legitimate authority to bring both leadership teams together to work on the integration process.

During this phase, both groups should retain their prior structures and continue to serve their respective customers as they did in the past. The maxim is, "Don't break anything!"

If groups are being moved into an existing organization, during this phase, they're given a safe home. Their temporary boss makes them comfortable in the new organization, familiarizes them with its culture, and deals with the necessary administrative changes (such as HR issues).

In a consolidation of a decentralized function, there's another issue to deal with in this phase. A shared-services organization receives resources (people, assets, and budgets) from business units in trade for a promise to deliver services. There are two dangers here:

One is that business units will retain resources that should be consolidated. Clear principles should be crafted to define what's in scope and what's not. Principles should also define what to do with "rainbow" individuals who are only partially in scope (and part of their work remains with the business unit).

The other danger is that business units will expect more than is reasonable; they may hand over resources sufficient to build a Chevrolet, and then demand Rolls-Royce service in return. Thus, it's critical to document current service levels being delivered by the decentralized groups before consolidation.

Integration Phase

The Integration phase is when the groups are merged into a single, integrated structure. This is a delicate process, with risks such as:

- Confusion among clients and staff, and missed commitments
- Morale problems, and undesirable turnover
- Winners and losers, instead of gleaning the best of both sides
- A structure that is simply an enlarged version of the old organization, without real synergies

If the receiving organization has already applied the science of structure, then the challenge is to merge the new group into an already carefully engineered organization. A Rainbow Analysis of the incoming group provides the needed map.

However, in other cases, there's an opportunity to design an entirely new, integrated structure. The implementation process described in Chapter 45 can be used, just as it's applied to the reorganization of a single organization. It provides a fair, fact-based, participative process; treats all staff equitably; and produces a new structure that gleans the best from both sides.

The starting point is a Rainbow Analysis of both organizations. Using the common language of the Building Blocks, participants identify where there are two or more groups in the same line of business. It shows groups in multiple lines of business (rainbows), which perhaps should be divided among separate groups in the new organization. And it may show gaps in both organizations which should be filled in the new structure.

In either scenario, participation of all the affected managers is

Chapter 47: Consolidation of Shared Services and Acquisition Integration 349

essential. This brings to the table all the knowledge that's needed to really understand both organizations. Furthermore, the process itself is a powerful team-building experience that breaks down walls, builds working relationships, and generates commitment to the new organization — all while accomplishing real work, the design of the new structure.

When it comes time to select leaders, the process should be fair and place the best qualified people where they're most needed, regardless of which group they came from. HR policies should protect managers who are asked to drop down a level, maintaining their compensation for a period of time (as per the ground-rules).

Even if headcount savings are expected, it's important to go through the Integration phase without cuts. No productivity improvements have been delivered yet. And it's essential to build a safe environment for participation and a positive outlook on the new organization. Any necessary cuts occur in the next phase.

Throughout this phase as well, staff continue to perform their existing work for their existing customers, so as not to miss any commitments.

Optimization Phase

Once the new structure is up and running, with similar lines of business from both groups consolidated under a single manager, this final phase is when the cost savings and synergies are delivered.

In Optimization, within each group, each manager implements process improvements, benefiting from the best practices in both organizations (the pieces now together under him/her).

Staff's assignments may be adjusted to put the right talent on each job, and to increase specialization. And redundancies may be eliminated.

Vendor contracts may be consolidated. And redundant assets may be written off.

And when similar services are put side by side in the catalog of an integrated group, customers may see opportunities for cost savings and collaboration through shared solutions.

As managers optimize their groups, the promised cost savings and synergies are realized over time.

SYNOPSIS

» Consolidations of shared-services functions and the integration of mergers and acquisitions are similar processes.

» Successful integrations exhibit a number of benchmarks, including equitable treatment of people, involvement, and change management.

» There are three distinct phases: Consolidate, Integrate, and Optimize. Don't expect the benefits until the Optimization phase.

» The process of restructuring (Chapter 45) is a good way to design and implement an integrated structure.

~ PART 10 ~
Leadership Questions

> *The first responsibility of a leader is to define reality.*
> *The last is to say thank you.*
> *In between the two, the leader must become a servant....*
>
> Max De Pree [97]

This final Part helps executives think through their role and their plans with regard to the structure of their organizations.

Its three Chapters answer three questions that you might ask:

Chapter 48: Should I do this? Do the benefits justify the costs?

Chapter 49: When? Where does structure fit in a broader transformation strategy?

Chapter 50: What should I do next? What practical steps should I take to start down this path?

Chapter 48:
Should I Do This?
Benefits That Justify the Costs

Restructuring an organization isn't easy.

Sure, it's easy to simply draw boxes and assign names. But as most of us have experienced, doing so does little to improve an organization's effectiveness.

There's a reason why executives take such ineffective short-cuts. Building a high-performance organization takes meticulous planning. And a transformational, participative process adds to leaders' workloads. Plus, the process is stressful, since people worry about their careers, their status, and their power.

What, then, are the benefits that justify these costs?

A principle-based structure delivers both efficiencies and effectiveness. And it transforms an organization's culture as well.

Efficiencies

Efficiencies come from a variety of sources:

- **Overlaps are eliminated.** Time is saved by eliminating redundant efforts, by eliminating territorial disputes, and by reducing the time that managers spend sorting out who does what.

- **Role clarity permits greater span.** With clear domains, groups are self-directed to a far greater extent. This permits a flatter organization, since it reduces supervisory workloads.

- **Professional collaboration is encouraged.** Bringing together related specialties induces more professional exchange. This precludes redundant research, accelerates organizational learning, improves productivity, and enables more reusable components instead of reinvention.

Efficiencies reduce rates (the cost per unit of work); the organization's products and services are a better deal. But be cautious about using efficiency gains to justify a restructuring initiative. This may lead to expectations of reduced budgets and headcount.

In practice, efficiencies are often reinvested in doing more with the same budget, not the same with less budget. Organizations catch up on critical sustainment tasks like training, innovation, planning, and relationship building. They improve the quality of their services and reduce risks. They catch up on deferred maintenance. They're more responsive to their customers, addressing backlogs and pent-up demand. And they find new opportunities to deliver more value by filling gaps.

Also, promising cost savings from a restructuring associates the new organization with job losses. This can set staff against the change, making it tough to realize its potential benefits.

And remember, efficiencies develop gradually over time as the new structure settles into place, not instantly upon announcement of a new organization chart.

For every minute spent organizing, an hour is earned.

Benjamin Franklin

Effectiveness

Effectiveness is more important than efficiency, even in companies concerned about costs. Efficiencies produce marginal cost savings, whereas improving effectiveness enhances an organization's contribution to the business — a highly-leveraged benefit.

For example, reducing headcount by one might save $100,000 per year. But if greater effectiveness means one more strategic project each year, the benefits could be counted in the millions.

There are many reasons why a well-designed structure improves effectiveness:

- **Gaps are eliminated.** Assigning accountability for missing lines of business can deliver value that wasn't available from the old organization (even if they're not full-time jobs).

    If Engineers or Service Providers were missing, filling the gaps delivers new products and services.

    Creating a new Sales function in an internal service provider improves relationships and strategic alignment, which should lead to an ongoing stream of very high-payoff projects.

    If Coordinator functions were missing, filling gaps improves product architecture and standardization, planning, policies, and safety (security, compliance, and risk).

- **Specialization increases.** As scattered campuses are consolidated, and by choosing the right basis for substructure, impossibly diverse jobs are replaced with well-focused specialties. Staff become more competent, and conflicts of interests are eliminated.

Greater focus and specialization produces improved productivity, faster and more reliable delivery, higher quality, lower risk, more innovation, less stress, and greater staff motivation.

- **Accountabilities for results are clear.** Groups are defined as lines of business, accountable for delivering products and services. This results-orientation delivers *results*.

   It also helps clients and internal customers, since it's easy to know where to go for one's needs.

   Precise domains also facilitate performance management. It's easier to hold staff accountable for results-based metrics.

   And people feel good about their jobs when they see the value of their work to customers.

- **Teamwork is enhanced.** Teamwork isn't limited to a few visible projects where executives assemble teams. It happens all the time, whenever any group gets help from its peers.

   Better teamwork improves performance. Instead of each group attempting to be self-sufficient, the best talent from a wide range of relevant specialties contributes to each project. This produces far better results than generalists working alone.

- **The organization is scalable.** A healthy structure makes it easier for an organization to grow, to take on new technologies and new missions, and to integrate acquisitions and consolidations. It's not only clear where new opportunities fit; it's clear who's accountable for discovering them.

   And for internal service providers, changes in clients' structure or strategies have minimal impact on the organization's structure (outside of Account Sales).

Thus, it's entirely possible to design a structure that lasts a lifetime, evolving naturally as new opportunities arise without the need for another major restructuring.

- **The executive bottleneck is eliminated.** Empowered staff don't need to be micro-managed. So executives can rise above the day-to-day and focus on more strategic challenges.

Cultural Transformation

In addition to improved efficiency and effectiveness, a healthy structure has powerful impacts on an organization's culture:

- **Empowerment:** When jobs are defined by what people "sell" (results), accountabilities are well defined; so staff can be empowered to produce those results as they see fit.

    Empowerment improves reliable delivery, since people have the resources and authorities they need to get the job done.

    Empowerment encourages creativity and innovation, as staff think about the best ways to accomplish the results they've agreed to deliver.

    Empowerment also encourages quality, since staff are fully responsible for their own results. People know they have to maintain what they sell; so there's an incentive to do things right the first time, rather than wasting time and money correcting mistakes later. This core principle of Total Quality Management is imbedded into the fabric of the organization.

- **Customer focus:** Running a businesses within a business, staff recognize they're funded (given a budget) to serve customers, both within and outside the organization.

They focus on delivering products and services to customers, rather than on bureaucratic territories, procedures, and tasks.

They understand what their customers want by listening, and by responding to customers' priorities (not their own).

And they build effective working relationships with customers, and are pleasant to do business with.

- **Entrepreneurship:** Since groups are defined as lines of business, everybody is an entrepreneur.

    Of course, entrepreneurs know they have to be reliable to stay in business. They learn not to make commitments they can't keep, and to keep every commitment.

    Entrepreneurs have to remain competitive, which encourages frugality. They manage their costs, and seek to minimize their overhead rather than maximize their empires. And they propose "buy" over "make" if it's more economical, making the best use of outsourcing vendors.

    Entrepreneurs continually improve their tools, methods, and processes as they strive to be customers' supplier of choice.

    And entrepreneurs are innovative, since they're responsible for the future of their businesses. They explore new products and services, track emerging technologies, and acquire new skills. They do what it takes to enhance their businesses by delivering more value, now and in the future.

Great leaders know that the benefits of cultural change are *not* intangible; they directly improve performance. For example, Mary Barra, CEO of General Motors, was asked to list the most important changes she made to prepare a slow-moving, recall-plagued company for the future in a rapidly changing industry. She said: [98]

I would say... it was just a commitment to two things. One: To <u>focus on the customer</u> and to put the customer in the center of every decision that we make.... Two: ... We <u>empowered</u> our technical team and our chief engineers to do award-winning products defined by the customer.

Case Example: Transformational Benefits

Although a healthy structure is not a comprehensive treatment of culture, its many impacts on culture are profound. Here's an example: [99]

Marcy was a business process expert in a growing food packaging company. Smart, dedicated, and capable, she knew how to fix processes like the ones causing the company to miss shipments.

The Chief Operating Officer (COO) respected her abilities, and asked her to fix a struggling plant that reported to him. Marcy mapped the plant's current processes, found the bottlenecks, and redesigned the process from raw materials to finished products.

But, then she ran into a brick wall. No matter how "right" her recommendations were, they'd be of little value unless the staff in the plant embraced and implemented them. But plant staff resented her, and complaints worked their way up the chain.

Her boss pulled her off the project, and her recommendations weren't implemented.

The problem wasn't any lack of interpersonal skills. It was

Marcy's misunderstanding of what business she was in, and whom her customers were.

She thought she owed the COO a solution to the problems in the plant. But she couldn't sell him that because she didn't have the authority to change the way other people worked.

Furthermore, she wasn't accountable for making those shipments. Plant managers were; and they were resistant to ceding authority over how they work to Marcy (rightfully, as per the Golden Rule).

When Marcy started to think of herself a business within a business — a one-person consulting company — her perspective changed. She understood she couldn't sell "changing others" to the COO. The most she could do was to sell a process-facilitation service.

It quickly became clear that her customer had to be the plant staff, not the COO. To get the dynamics right, the COO had to tell the plant manager that <u>he</u> was responsible for improving throughput, and then offer Marcy's services to help him achieve that objective.

By defining her job as a consulting business, and by adopting a culture of entrepreneurship and customer focus, Marcy was positioned for success.

The benefits weren't limited to the value of Marcy's time. Her effectiveness contributed directly to the entire company's success.

Attracting a New Breed of Employees

One goal of a good leader is to become the "supplier of choice" to customers. But an equally important goal is to become the "employer of choice" to staff. A healthy organization attracts, motivates, develops, and retains the best people.

This isn't easy. The expectations of the workforce are changing. Millennials don't plan to join a firm for life. They think more like independent contractors, in control of their careers and lives.

There's no reason to believe this is a temporary phenomenon, characteristic of only this one generation. Ade McCormack sees this as a return to our intrinsic nature as a species. *100*

McCormack reflects on our early hunter-gatherer stage of evolution. Survival required teamwork; we hunted and gathered in packs. We were rewarded for results, not effort, and saw the value of our work ("eat what you kill"). And we had to adapt to a changing environment, requiring creativity and agility.

These same traits were rewarded in the agricultural era, with even greater emphasis on collaboration through trade. And the need for planning arose, since crops and livestock required effort well before they bore fruit.

McCormack concludes that our natural desire is to be social, results-oriented, creative, and to plan our own futures.

The industrial revolution changed that. Factory jobs emphasized routine over creativity, because independent initiative disrupted tightly controlled assembly lines. People became isolated, working around others but without having to collaborate to do their jobs; in fact, chit-chat was forbidden. And doing tasks on an assembly line, people were a long way away from seeing the value of their labor to others.

This foreign (to our nature) environment carried over into office work, where we're assigned tasks and often disempowered. Counter to our nature, we're rewarded for being compliant and reliable, not creative and results-oriented.

McCormack observes that as the industrial era fades into the memory of past generations, the workforce (such as millennials) demands jobs that are more gratifying — involving independence, diverse challenges, collaboration, creativity, and rewards connected with accomplishments — the very traits that were ingrained in us since our hunter-gatherer days.

Principle-based organizations provide exactly that. People are empowered as entrepreneurs. They can (indeed must) take initiatives and be creative. They work closely with others on teams (as specialists must). And being in business to produce products and services for others, they see the fruits of their labor; they feel a sense of accomplishment in producing results, and a sense of self-worth in delivering value to customers.

Healthy structure provides an environment that attracts, motivates, develops, and retains the new breed of workers.

Bottom Line

In summary, **an investment in your organization's structure is justified by the increased value the organization delivers in every aspect of its work.**

It delivers better responsiveness, speed, cost, reliability, quality, product integration, flexibility, creativity, innovation, customer focus, and alignment with the needs of its customers — the kind of organization customers want to do business with.

At the same time, it engages every bright mind, empowers staff to contribute all they can, and helps everyone succeed — the kind of organization people want to work for.

A healthy structure can make an organization both a *supplier of choice* to its customers, and an *employer of choice* to its staff.

> *I'd have to say that a principle-based structure has been one of the best investments I've made.*
>
> Sergio Paiz, CEO, PDC

SYNOPSIS

» Efficiencies result from eliminating overlaps, increasing span of control, and improving professional collaboration.

» Effectiveness results from elimination of gaps, increased specialization, clear accountabilities, enhanced teamwork, a scalable structure, and eliminating the executive bottleneck.

» Structural change can induce a change in culture, to one of empowerment, customer focus, and entrepreneurship.

» In addition to benefitting an organization's customers as their supplier of choice, a healthy structure attracts and retains the best talent (and the new breed of employees) as an employer of choice.

Chapter 49:
When?
Place Within a Transformation Strategy

As powerful as structure is, a caveat is in order: It's not a panacea that will solve all your problems. Structure is one of **five organizational systems**, listed in Figure 28, that make up **the ecosystem in which we work**. *101*

Figure 28: Five Organizational Systems

- **Culture:** the widely held beliefs and patterns of behaviors (habits and conventions)
- **Structure:** the definition of jobs and the reporting hierarchy (organization chart), as well as the processes that combine people into teams across organizational boundaries
- **Internal economy:** the collection of processes that manages resources, including budgeting, rates, demand management (priority setting), commitment management, time management, accounting, chargebacks (if any), and financial reporting
- **Methods and tools:** the procedures, methods, skills, and tools that augment staff's capability to do their jobs (distinct from cross-boundary workflows which are determined by structure)
- **Metrics and rewards:** dashboards and key performance indicators (KPIs) that inform people about how they are doing, and incentives for improving performance on those metrics

These five organizational systems work in concert to send signals that guide people in their work. They shape the character of an organization, and determine whether the organizational environment is productive and effective.

This Chapter helps you formulate a broader transformation strategy, and determine where structure fits in that plan.

Be Aware of All Five

Even if you're only interested in structure right now, recognizing the presence of all five organizational systems is important for this reason: If you use structure to solve problems whose root causes are other organizational systems, you're likely to create destructive side-effects, and you probably won't solve the real problems.

Let's say, for example, that your organization must be highly responsive to quick-turnaround projects. And yet you also have to do large, complex projects. This is essentially a resource-allocation challenge — a matter of assigning some of your resources to big projects and holding some aside for quick wins.

But if you try to address this challenge in structure instead of the internal economy, you'll create two groups: a rapid-response group, and a traditional projects group. Chapter 25, "Quick Versus Slow," describes the many serious problems this creates, not the least of which is "two classes of citizenship" with internal rivalries, higher costs, lower quality, and demotivated staff.

Or, as in Chapter 23, you might mistakenly set up a disempowering "innovation" group to reserve time for planning and exploring, rather than giving everybody time for thinking about the future of their own entrepreneurships.

Part 5 includes many more examples. Again and again, the lesson is: Don't use structure to solve a problem that's rooted in another organizational system.

Sequence Within an Organizational Strategy

Of course, if structure is the root cause of your most pressing concerns, then it's the right place to start.

But will the structure you design work as well as intended without changes in some other organizational systems?

And what if you'd like to address all five systems in a comprehensive transformation strategy.... What's the right sequence?

Each system reinforces the others, and makes the others easier to implement. This is especially true if all are working toward the same vision: the business-within-a-business paradigm.

There are some interdependencies that are worth noting as you consider what other changes you might make after a restructuring, or sequence the five systems into your transformation strategy.

Culture

Contrary to popular opinion, culture is one of the easier organizational systems to change. [102] The key is focusing on behaviors rather than values. In an effective culture change process, a leadership team crafts a set of behavioral principles, and then teaches them, models them, and measures compliance. This approach can transform an organization's culture in less than a year.

Changing culture has significant benefits, and amplifies the benefits of structure. For example, the benefits of a business-within-a-business structure are enhanced when culture teaches people the specific behaviors of empowerment and entrepreneurship.

Arguing for culture before structure, a culture of teamwork may reduce leaders' tendency to demand independent silos.

On the other hand, some aspects of culture cannot be implemented until the structure is right. For example, people can't be customer focused if their jobs are defined by tasks (e.g., roles and responsibilities) rather than lines of business, making it hard to know whom their customers are.

And if structure makes people accountable for doing tasks rather than delivering results, they can't know what they need to buy from other groups. Someone at a higher level, who is accountable for deliverables, must design processes and form teams. In such a structure, preaching a culture of teamwork may not accomplish much.

With arguments both ways, the order isn't critical. You can base your strategy on which system is most urgent.

But if you treat structure first, remember that you may have to be patient and wait until the culture comes into alignment before you see all the benefits of the new structure.

You may wish to establish just a few cultural principles which will reinforce the new structure, in particular those which encourage cooperation as one organization, teamwork, empowerment, customer focus, and entrepreneurship. If you do, remember to focus on actionable behaviors, not values and feelings.

(Some examples of the behaviors that support a principle-based structure are described in the Supplement to this book; contact <ndma@ndma.com>.)

Internal Economy

The "internal economy" is the collection of processes that manages resources (time and money), including: budgeting, rates, demand management (priority setting), commitment management, time management, accounting, chargebacks (if any), and financial reporting. [103]

An effective internal economy is essential to an organization's success. For example, performance is undermined if expectations exceed resources, causing staff to attempt to satisfy unreasonable demands by sacrificing time for training, relationship building, process improvements, and innovation. And teamwork is undermined if priorities are set by individual managers for their own groups, rather than by clients for the entire organization, or if everybody is overworked and has no time to help peers.

A healthy internal economy makes it easier to fix the structure. At a minimum, it rescues staff from the overwhelming pressures of expectations beyond resources, giving them time to invest in structure and other transformation initiatives. And a common set of priorities is essential to teamwork.

Conversely, a healthy structure makes fixing the internal economy easier. When lines of business are well defined, it's much easier to develop a catalog of products and services, and to calculate product/service costs. And it's easier to know the cost of projects and services when walk-throughs tell you who's on each team.

Again, the sequence can be driven by which problems are most urgent. And again, you may not see all the benefits of a new structure until problems in your internal economy are addressed.

Methods and Tools

Introducing new methods and tools enhances the competencies within professions.

Some methods and tools are relevant to many specialties, such as project management and time management. These can be introduced at any time.

But profession-specific methods and tools are best introduced *after* structure. If there's no home in the structure for a given specialty, its methods and tools won't be sustainable. Conversely, an entrepreneurial structure creates a natural hunger for methods and tools to improve one's performance.

Also, note that methods and tools are rarely transformational. They fine-tune the performance of an organization, but they don't change its paradigm.

So generally, methods and tools are placed late in a transformation strategy, well *after* structure.

Metrics and Rewards

Metrics (KPIs) fall in two categories: dashboards that help people do their jobs, and performance metrics. Dashboards have a greater impact on performance. But performance metrics are tied to rewards, so they get more attention.

By the way, to support empowerment, metrics of *how,* rather than *what,* belong on dashboards delivered to the people doing the work, but should never be part of performance evaluations.

There are many reasons why both types of metrics should come *after* structure. For one, structure clarifies each group's unique objectives, providing a basis for designing relevant metrics.

Furthermore, metrics done too soon can be harmful. If the structure includes overlapping territories, metrics and rewards only serve to motivate people to fight even harder with one another. And if jobs have inherent conflicts of interests, metrics and rewards only tear people apart more quickly.

Also, metrics may cause resistance to transformational change, since staff may feel that change interferes with their ability (in the short term) to achieve their metrics.

There's an old saying: "If you want to fix it, measure it." But this is said by people pursuing incremental improvements within existing organizations. Metrics fine-tune behaviors and performance, but they're not transformational.

For all these reasons, metrics and rewards are best addressed toward the end of a transformation strategy.

---

*Not everything that can be counted counts,
and not everything that counts can be counted.*

Albert Einstein

Transformation Action Plan

Ultimately, all five systems must be aligned to achieve a truly healthy, high-performance organization. Transformation isn't a short-term project. Just as every good executive has ongoing business and technology strategies, great leaders continually focus on their organizational strategies.

While there's no one sequence that's right for every organization, note that structure is the most powerful of the five. Without a healthy structure, cultural principles will be seen as empty exhortations; the internal economy will incite greater friction as entrepreneurs compete for business; methods will not find a home; and metrics will exacerbate unclear and conflicting objectives.

For many organizations, structure is the right place to start.

SYNOPSIS

- » An organization's "ecosystem" comprises five systems: culture, structure, the internal economy, methods and tools, and metrics and rewards.

- » Three systems are transformational and should come early in a leader's strategy: culture, structure, and the internal economy.

- » Each system complements the others; there's no "right" order. But be aware of this: Whichever you treat first, some of the expected benefits may await treatment of the others.

- » While ultimately all must work in concert, for many leaders, structure is the best place to start.

Chapter 50: What Should I Do Next? Ideas to Action

I believe your organization, like every organization, is made of good people.

Sure, not everyone is a super-achiever. But your organization is filled with people who are sincere about its mission and their own responsibilities; who work hard; who freely offer their diverse talents and creative perspectives; who get along with one another; and who want to do some good for others and leave the world a better place. You have the raw material that makes organizations great.

But when we expect more of our staff than they can deliver; isolate them in silos; pit them against one another; disempower them by holding them accountable for things they can't control; disrespect them with micromanagement — we not only undermine their effectiveness. We fail to serve are clients; our stakeholders in the community; our shareholders/taxpayers/donors; and the people and families who trust us to provide their livings and their careers.

Building an effective organization is one of the most important responsibilities of leadership. It's more important than short-term challenges like coming up with the right strategies, delivering important projects, or closing the next big sale.

Great leaders leave a legacy of highly effective organizations that are designed to succeed forevermore.

This book gives you the engineering science, and a proven process of change. You know how to:

- Make incremental changes ("tweaks") that move your organization toward a consistent evolutionary vision, and solve pressing structural problems without creating new ones.

- Or starting with a clean sheet of paper, design a really great structure that will last.

- Safely invite managers to participate in the design process.

- Communicate the various roles within the organization clearly to clients and to your staff.

- Establish practices of high-performance teamwork.

I know structural change isn't easy.

One reason is that it's emotional. Reorganizations affect people's livelihoods, sense of competence, and power base. People have a natural tendency to protect the status quo.

Another reason it's hard is because everybody has an opinion, and there's no way to appease all those points of view. Your challenge is to marshal all those diverse views into a consensus.

And when outsiders make demands, you have to be tough, and structure your organization to deliver what customers ask of it, not just comply with their opinions about your structure.

And then there's the issue of time. You may feel that your staff are too busy to take this on. But will there ever be a time when you're not too busy?

In a small monograph on time management (tinted with religion) titled "Tyranny of the Urgent," Charles Hummel delivered a profound message: [104]

We live in constant tension between the urgent and the important. The problem is that many important tasks need not be done today.... But often... less important tasks call for immediate response — endless demands pressure every waking hour. With a sense of loss,... we realize we've become slaves to the tyranny of the urgent.

Structure may or may not be urgent, but it's certainly important.

To step above all the obstacles, and to bring transformational change to your organization, requires real leadership. You have the tools: the science and the implementation process.

And you know an easy way to start: the Rainbow Analysis.

I hope this book helps you look past the pressures of the day, and create your legacy as a great organizational leader.

SYNOPSIS

» Building effective organizations is one of the most important responsibilities of leadership. It's more important than any short-term challenges like making the right decision on strategies, delivering projects, or closing the next big sale.

» Great leaders leave a legacy of highly effective organizations that are designed to succeed forevermore.

» The Rainbow Analysis is an easy way to start.

~ APPENDIXES ~

Three Appendixes support your reading of this book. They are:

Appendix 1: Terminology

Appendix 2: Supervisory Duties

Appendix 3: Theoretical Underpinnings

These are followed by the Endnotes (references) cited throughout the book, and finally the Index.

Appendix 1:
Terminology

> *Clear language is the basis for clear thinking.*
>
> N. Dean Meyer

This Appendix defines key terms used throughout this book, and establishes a common language for discussions of structure.

Organization

An "organization" is any collection of people working within a common entity (e.g., within the same company) and sharing a purpose (a mission).

It may be a corporation or company, a government agency, a family business, a volunteer association, or a union.

An organization may seek to make a profit, or it may be not-for-profit.

Typically, all the members of an organization work for a common boss (the "top executive"), although this is not always the case; the organization may be a confederation where its members have banded together for a common cause (such as a club).

Organizations exist within organizations. An organization may also be a department within any of these enterprises. The scope of the organization is a matter of your purview. You might use the word "organization" to mean all the people at all levels who report to you.

- If you're the CEO of a multinational corporation, your organization may include numerous operating companies as well as corporate departments.
- If you're the president of a company within that conglomerate, the relevant organization is your company.
- If you're an executive in charge of an internal service provider within that company, the organization you care most about is your department.

Enterprise

I use the term "enterprise" to mean an entire corporation, not-for-profit entity, or government entity (one that's intended to be mostly self-sufficient).

An enterprise is itself an organization, and it's made of many sub-organizations.

Internal Service Provider

Internal service providers are organizations whose primary mission is to provide products and services to others within the same enterprise.

They include traditional support functions like finance, human resources, administration, and information technologies. They also include functions like engineering, manufacturing, customer service, sales, and marketing, all of which serve the various brand managers who are accountable for the company's product lines.

Organizational Structure

Organizational structure defines the layout of jobs within an organization, and which jobs report to which managers (the reporting hierarchy). An organization chart defines **people's specialties**, and determines who is accountable for what.

Of course, changes in an organization chart inevitably change workflows. Therefore, organizational structure also defines people's relationships with one another, and how they collaborate to get work done — the **workflows** across structural boundaries.

Figure 29: Two Components of Organizational Structure

- Organization charts that define people's specialties and the reporting hierarchy
- Workflows that assemble the right specialists from throughout the organization on project and service-delivery teams

Group versus Team

A **group** is a collection of people with a common boss. A group is an organization, from the perspective of the manager of that group.

Groups exist within the structure of an organization, and appear on its organization chart. They are permanent, or at least intended to survive beyond a single project.

Teams cut across the structure, and should not appear on an organization chart. They're temporary associations of people who report to one or more groups, and who agree to work together to achieve a specific deliverable.

A team disbands at the end of its project or service, while its members continue to work for their respective groups.

Self-managed and Self-directed

A **self-managed group** is a group with staff but no supervisor. The staff in the group share the responsibilities of a supervisor. This is discussed in Chapter 34.

A **self-directed group** is a group with a supervisor (as is normal) where the staff are empowered and accountable. It does not require a great deal of direction from the manager. This term just describes a well-managed organization.

A **self-managed team** is quite different. It's a team, not a group, that accomplishes its mission without a great deal of intervention from its members' managers. The mechanisms that make this possible are discussed in Part 8.

Clients versus Customers

I use the word "client" differently from "customer."

Organizations exist to serve clients.

Clients are people outside the organization (except in the case of clubs where members are both staff and clients). Clients may be customers who consume its products and services; but at times these same people may also be suppliers to the organization.

If the organization is a company, its clients are customers external to the company. If the organization is an internal service provider within a company, its clients are, for the most part, others within the enterprise.

Customers are those who consume a group's products and services. They may be clients or others within the organization.

"Customer" and "supplier" refer to a relationship (rather than a group of people). They're applied in the context of a specific contract, where a customer "buys" something from a supplier (whether or not money changes hands).

Another group within your organization may buy your products and services, either for its own use or to add value and then pass them along to another customer. They are a customer to you; but since they're within your organization, they're not a client.

Here's an illustration of all these relationships (see Figure 30):

Client A (outside your organization) buys something from group B within your organization. In this relationship, the client is the customer, and group B is the supplier.

Group B in turn gets help from group C, also within your organization. In this relationship, group B is now the customer and group C is the supplier.

Group B also needs help from client D to do its job. In this relationship, group B is the customer and client D is the supplier.

Appendix 1: Terminology 381

Figure 30: "Customer" Versus "Client"

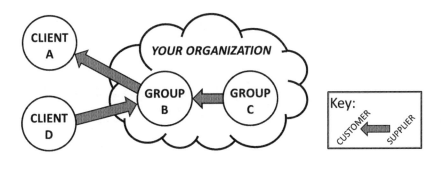

Broker versus Sell

A **broker** helps a customer and supplier come to an agreement. For example, the Sales function may help a client buy a solution from an Engineer within the organization.

But brokers aren't accountable for the delivery of products/services (as if they were the prime contractor). They just sell the brokerage service.

Consider real estate brokers. They don't take possession of a house, and then resell it. And they're not accountable for the quality of the house. (The seller is.) This is a brokerage service.

This can be confusing because, in the English language, "sell" has two meanings: provide, and broker. Throughout this book, I use the word **sell** strictly to mean "provide," i.e., accountability for the delivery of the product/service.

The Sales Building Block sells a brokerage service. Engineers sell products. Service Providers sell services.

Appendix 2:
Supervisory Duties

Regardless of grade level, anyone who manages a group of people or a distinct line of business is a "supervisor."

There are some duties that all supervisors (in all functions) do.

Some of these have to do with the content of the group's work, termed "domain-specific." Others are "generic," and don't require in-depth knowledge of the function.

Domain-specific supervisory duties address the content of the group's business, including the following:

- Business strategy planning (for your own business within a business)
- Budget planning and financial management
- Product management of your group's products/services, including product definitions, pricing, and product strategies
- Marketing, building customers' awareness of the availability and value of your products/services
- Sales, proposal writing
- Commitment management (contracting)
- Acquiring feedback from customers and suppliers
- Determination of subordinate organizational structure; coordi-

Appendix 2: Supervisory Duties

nation of subordinate domains; integration of consolidated groups; arbitration of disputes

- Hiring staff (employees, contractors) and vendors
- Coaching, development planning, career counseling, and motivating staff
- Performance management (staff's objectives and appraisals, discipline)
- Allocation of rewards
- Facilitation of decision-making within the group, and making decisions when consensus is lacking
- Managing assets owned by the group, inventory management, and capacity management
- Acquiring methods and tools, establishing professional practices, and continually improving processes
- Acquiring needed services and supplies from other groups (subcontracting), including all needed internal support services
- Management reporting, and representing the group to peers and superiors
- Ensuring compliance with all applicable laws, regulations, policies, standards, and enterprise and organizational rules and mandates
- Business continuity planning for the group
- Security, compliance, and safety of the group, and of its products and services

Generic supervisory duties do not require in-depth knowledge of the group's business. They include the following:

- Resource management; work allocation; scheduling of work
- Discipline (after performance management determines the need for it)
- Transition facilitation, helping staff with the personal issues that naturally occur during change processes
- Official business, such as signatures and approvals
- Location-specific policies such as hours of operation, dress codes, and security access rights

Domain-specific supervisory duties must be done by someone in the group — either its manager, or a representative of a self-managed group (Chapter 34).

Generic supervisory duties can be delegated to others. For example, for geographically distant staff, managers may wish to delegate to a "site coordinator" (described in Chapter 36) some or all of the generic supervisory duties, but not the domain-specific duties.

Appendix 3:
Theoretical Underpinnings

This Appendix is an overview of the theoretical underpinnings of the science of structure.

Organizations as Cybernetic Systems

The science of structure is based on the science of cybernetics — the study of the control mechanisms in dynamic systems. It's basic systems science, put to work.

Stafford Beer, the renowned British cybernetician, pioneered the application of systems science to organizational design. Beer defines a "system" as:

"...a group of elements dynamically related in time according to some coherent pattern." [105]

This describes organizations perfectly. **Organizations are a set of parallel, but interlinked, dynamic systems** — a set of ongoing functions, each controlled by various mechanisms, all operating in parallel, and yet coordinated so as to work together.

If we look more closely at Beer's definition of a system, we can make two inferences:

- His reference to *elements* implies distinct boundaries. Those elements (groups within an organization) must be clearly bounded (e.g., in terms of authorities, accountabilities, and resources), and individually effective at their purposes.

In a principle-based organizational structure, the various groups within it are clearly bounded (precise domains) and individually effective (empowered entrepreneurships).

- The elements are *dynamically related in a coherent pattern*. This suggests that explicit mechanisms are put in place to coordinate collaboration and ensure that people work synergistically. It also says that patterns of collaboration are dynamic — they adapt to an ever-changing environment.

In a principle-based organizational structure, the formation of teams is not left to chance, and roles within teams are well defined. The meta-process of teamwork dynamically interlinks groups in coherent pattern of high-performance teamwork, tailored to the needs of each unique project or service.

In applying cybernetics to organizations, two key principles are particularly relevant:

- The cybernetic principle of *feedback loops* — the signals that guide people in their daily work and decision making.

- The cybernetic principle of *requisite variety* — the ability of every element in the organization to adequately handle all the complexity and volume of work that it faces.

Feedback Loops

One of the most basic tenets of cybernetics is that systems are controlled by "feedback loops," signals that adjust systems to changing conditions.

Feedback loops are widely used for the control of machinery. Delightfully, one of the earliest practical applications was a self-

Appendix 3: Theoretical Underpinnings 387

filling wine bowl invented in Egypt in approximately 300 AD. A simple modern-day example is the thermostat in your home which turns on the heater when it gets too cold.

Organizations produce signals that guide people's performance. In organizations, this key cybernetic concept of feedback loops can be described in a simple truth: **People generally do what you reward them for doing.**

For example, you want staff to be frugal. But if your HR policies decide job grades and compensation based on the size of one's budget and headcount, then people are, essentially, paid to build empires.

This feedback loop is so powerful that managers will find ways to maximize their own best interests despite budget processes, plans, and management oversight. For example, they'll attempt to expand their headcount (grow independent silos) rather than team with peers, so as to maximize their job grades.

Organizational structure is a fundamental source of feedback loops. Job descriptions tell staff what they're supposed to be good at, and what they're accountable for. This is a powerful driver in itself, since people focus on the mission of their groups (their businesses within a business). And it's the basis for defining metrics and rewards (explicit feedback loops).

By the way, "reward" means much more than money; it's all the things that make people feel good, as per behavioral economics. [106] Most people are reasonably rational; but they optimize their *subjective* well-being, not money alone.

In the science of structure, performance is enhanced when monetary rewards are complemented with a sense of accomplishment in running a business within a business, and value in one's service to internal and external customers.

A healthy organization does not depend on altruism. It's unfair to ask people to sacrifice their best interests (and potentially those of their staff) for the sake of customers' well-being. It's also unreliable and fragile; altruism bets on people acting counter to the feedback loops, an unlikely proposition.

Rather than depend on altruism, a principle-based structure depends on "enlightened self-interests." [107] It aligns individuals' best-interests with the interests of the organization as a whole. With this alignment, maximizing one's own well-being *automatically* means doing what's right for the organization as a whole.

The key to this is that in a business-within-a-business organization, everybody's performance is judged by customers' satisfaction. Just as in the global economy, teamwork occurs because it's in individual's best interests to please customers and utilize suppliers. Thus, feedback loops align individuals' performance with their customers' needs, and ultimately with those of the enterprise as a whole.

When staff do what's right because that's what's best for them, then there's far less need for management oversight, disempowerment, rigid plans, shared metrics that people can't fully control, and audits. It permits a high degree of empowerment; people can act independently and the organizational system will automatically coordinate their activities. This frees leaders to focus their time and energy on more strategic challenges.

The power of this systems perspective is immense. A well-designed system of feedback loops allows individuals to excel, minimizes dysfunctional politics, achieves synergies, integrates results, and adapts and evolves within an ever-changing business environment.

Requisite Variety

A second key cybernetic principle that applies to organizational structure is W. Ross Ashby's "Law of Requisite Variety." [108]

"Variety" is complexity times pace (complexity times volume divided by time) — a measure of the information and events an organization must process each day.

$$VARIETY = COMPLEXITY \times PACE$$

An organization is confronted with tremendous variety. Think of all the knowledge within any one profession, and how rapidly it's evolving. Add the complexity of diverse customer demands. Multiply that by all the many disciplines involved in producing your organization's products and services. And multiply all that by the pace of events that affect you, things like:

- Economic volatility instantly ripples around the world.

- Changing political boundaries and alliances bring new opportunities as well as threats.

- The globalization of markets and competition means that competition is more intense, and events anywhere in the world can affect your organization.

- Strategic partnerships are bridging the boundaries between organizations, presenting new collaborative opportunities and challenges.

- Increasingly complex legal and regulatory restrictions are putting new, perhaps conflicting, demands on organizations.

- Better educated customers expect more of their suppliers. For example, customers are demanding customized service at the price of mass production, and individualized marketing ("segment of one").

- Technological complexity is exploding in every field, creating tremendous new strategic opportunities and threats (e.g., the digital enterprise).

- Sheer size (scale) adds to variety. As an organization grows, the number of possible collaborative relationships between people grows exponentially, further adding to the complexity of the work environment.

Ashby observed that *variety always balances.* In other words, if too much variety confronts an organization such that it exceeds its variety-processing ability, the incoming variety must somehow be reduced. If this is not done in a planned way, things will "fall through the cracks."

Since an executive cannot keep up with every detail going on in an organization, "filters" — such as subordinate layers of management and administrative assistants who screen emails and telephone calls — reduce variety on the way into the executive's office.

If filters are not well designed, they'll arise haphazardly. For example, if too many messages accumulate, some are bound to go unanswered. Despite an executive's best intentions, variety will balance. But the wrong messages may be ignored in this haphazard approach to balancing variety.

Similarly, since executives cannot tell everyone what to do every minute of the day, amplifiers — such as policy statements, administrative assistants, and layers of management — expand variety on the way out of an executive's office. They, too, are best designed consciously rather than left to chance and the rumor mill.

Principle 2, the value of specialization, is derived directly from the cybernetic concept of requisite variety. With our limited variety-processing abilities, individuals can only be excellent at one profession at a time.

Designing an organization chart is a matter of distributing the total variety-processing requirements of an organization among its groups. The purpose of the organization chart is to define those components, each with feasible requisite variety.

The Building Blocks are each well focused on single domains of expertise. Thus, a group dedicated to a single Building Block (not a "rainbow") has the most manageable requisite variety.

Additionally, empowerment reduces the requisite variety of leaders.

And the meta-process of teamwork reduces the requisite variety of project managers, as they subcontract for deliverables rather than attempt to manage the entire team's tasks.

The Science of Structure

The science of structure views organizations not as assembly lines of tasks, or a simple hierarchies of control, but rather as dynamic systems of interacting components, each responding to feedback from its environment, and at the same time interacting with other components as part of a coherent system.

A principle-based organizational structure addresses the requirements of an effective dynamic system, including feedback loops and requisite variety.

---

*The most important job of a leader is to
design an organization in which everyone can succeed,
with or without her.*

N. Dean Meyer

ENDNOTES
(References)

1. Hill, Linda, Greg Brandeau, Emily Truelove, and Kent Lineback. "Collective Genius." *Harvard Business Review*. June, 2014.

2. Miles and Snow described the three strategic vectors as "entrepreneurial, engineering, and administrative": Miles, Raymond E. and Charles C. Snow. *Organizational Strategy, Structure, and Process*. New York: McGraw-Hill. 1978.

3. I'm grateful to Steve Gelb for this case study.

4. Various "contingency theories" have attempted to explain why some structures work better in some environments than others. For example: Burns, Tom and George M. Stalker. *The Management of Innovation*. London: Tavistock. 1961.

   Contingency theories are akin to saying, "People who don't need to go very fast seem to choose low-powered cars." But that observation doesn't mean that the organizations they studied made the right choices. It's possible that a benign business environment meant that they just never had to invest in a high-performance machine. But if conditions were to change, these organizations would be poorly prepared.

   Note that a high-powered car can get very good gas mileage at low speeds. And when the need arises, you'll be able to go fast. The science of structure explains how to build an organization that performs well under any circumstances.

5. For more on organizations as systems, see Appendix 3.

6. In government, the dysfunction of separating authority over how work gets done from accountability for results is sometimes manifested as a separation of "policy" from "execution," where policy-makers decide how others do their jobs. As anyone on the execution side will tell you, this is fundamentally disempowering and unproductive.

7. Urwick, Lyndall. *The Elements of Administration*. New York, NY: Harper & Row. 1943.

8. I'm grateful to Walter Dong for this case study.

9. Mintzberg, Henry. *The Structuring of Organizations.* Englewood Cliffs, NJ: Prentice-Hall. 1979. Pages 163-168.

10. I'm grateful to Brian Letzkus for this additional example of a motive for a client-based structure.

11. Ashby, W. Ross. *Introduction to Cybernetics.* London: Chapman & Hall. 1956.

12. The term "T-shaped" was first used to describe technology-deep professionals with a broad awareness of business so as to be able to relate to IT's internal clients: Guest, David. "The Hunt Is On for the Renaissance Man of Computing." London: *The Independent.* September 17, 1991.
I generalized the term to apply to every profession.

13. Smith, Adam. *The Wealth of Nations.* 1776.

14. Mintzberg, Henry. *The Structuring of Organizations.* Englewood Cliffs, NJ: Prentice-Hall. 1979. Pages 73-75.

15. Fouraker and Stopford described an organization that optimized specialization, but lacked the necessary coordination and control processes inherent in treating managers as entrepreneurs (Principle 7), and the teamwork mechanisms described in Part 8. It didn't work well:
"...profit contribution of functional specialists could not be measured against performance" and *"...problems of conflict and coordination at the lowest levels of the organization would frequently have to be passed up to the highest functional levels for adjudication. And some operating issues could not be settled there, but would have to reach the office of the chief executive."*
Fouraker, Lawrence E. and John M. Stopford. "Organization Structure and the Multinational Strategy." *Administrative Science Quarterly.* Volume 13, June, 1968. Pages 47-64.
Not understanding how to induce high-performance teamwork across structural boundaries, they presumed that the management hierarchy was the primary coordinating mechanism and hence advocated silos for complex organizations, calling them "Type III" structures.

16. Frost, Robert. "Mending Wall." 1914.

17. Jacka, Mike and Paulette Jacka. *Business Process Mapping: Improving Customer Satisfaction.* John Wiley and Sons. 2009. Page 257.

ENDNOTES (References) 395

18. Mintzberg described six bases for substructure (which he called "groupings"): Mintzberg, Henry. *The Structuring of Organizations.* Englewood Cliffs, NJ: Prentice-Hall. 1979. Page 108-112.
    - Knowledge and skill
    - Work process and function
    - Time (shift)
    - Output (results)
    - Client (market)
    - Place (geography)

    The science of structure helps you make the right choices. The key is selecting a basis that matches the nature of the specialty of the function.

19. I'm grateful to Peter Bergeron of Simplex Wire and Cable Company for this case study.

20. Conway, Melvin E. "How Do Committees Invent?" *Datamation.* April, 1968.

    Research supporting the impacts of organizational communications patterns on product design was published as: MacCormack, Alan, John Rusnak, and Carliss Baldwin. "Exploring the Duality between Product and Organizational Architectures: A Test of the 'Mirroring' Hypothesis." *Harvard Business School.* Working Paper 08-039. 2007.

21. I'm grateful to Randy Roig for this case study.

22. Meyer, N. Dean. *An Introduction to the Business-Within-a-Business Paradigm.* Danbury, CT: NDMA Publishing. 2002.

23. Pinchot, Gifford. *Intrapreneuring.* New York, NY: Harper & Row. 1985.

24. Meyer, N. Dean. *Outsourcing: how to make vendors work for your shareholders.* Danbury, CT: NDMA Publishing. 1999.

25. Peter Drucker virtually equates entrepreneurship and innovation. See: Drucker, Peter F. *Innovation and Entrepreneurship.* New York: Harper & Row. 1985.

26. For those familiar with *Structural Cybernetics,* this Building Block was termed "Technologists."

27. For those familiar with *Structural Cybernetics,* this Building Block was termed "Service Bureaus."

28. For those familiar with *Structural Cybernetics,* this Building Block was termed "Machine-based Service Bureaus."

29. Another name for an operational planning process that produces a budget is "investment-based budgeting." See: Meyer, N. Dean. *Internal Market Economics: practical resource-governance processes based on principles we all believe in.* Danbury, CT: NDMA Publishing. 2013.

30. Meyer, N. Dean. *RoadMap: how to understand, diagnose, and fix your organization.* Danbury, CT: NDMA Publishing. 1998.

31. For those familiar with *Structural Cybernetics,* this Building Block was termed "Consultancy."

32. Some Chief Digital Officers do not report to IT, and may blur the line between clients and IT. Beyond the role of IT Sales (facilitation of the discovery of opportunities, and of the definition of technology requirements), they may actually take on accountability for technology-based business strategies. In doing so, they usurp authorities and accountabilities from peers such as Product Management and Marketing, violating the Golden Rule.

33. Brown, Ken and Ianthe Jeanne Dugan. "Arthur Andersen's Fall From Grace Is a Sad Tale of Greed and Miscues." *Wall Street Journal.* June 7, 2002.

34. Chandler, Alfred D. *Strategy and Structure.* Cambridge, MA: The MIT Press. 1962.

    Also: Drucker, Peter F. *The Practice of Management.* New York: Harper & Brothers. 1954

35. See, for example: March, James G. and Herbert Simon. *Organizations.* New York: McGraw-Hill. 1958.

    Also: Cyert, Richard and James G. March. *A Behavioral Theory of the Firm.* Englewood Cliffs, NJ: Prentice-Hall. 1963.

36. Treacy, Michael and Fred Wiersema. *The Discipline of Market Leaders: choose your customers, narrow your focus, dominate your market.* Reading, MA: Addison-Wesley Publishing Company. 1995.

    The concept of a firm's "core competency" was originally introduced by: Prahalad, C. K. and Gary Hamel. "The Core Competence of the Corporation." *Harvard Business Review.* Volume 68, number 3, 1990. Pages 79-91.

ENDNOTES (References)

37. Galbraith, Jay R. *Designing the Customer-centric Organization.* San Francisco, CA: Josey-Bass. 2005.

38. For internal service providers, outsourcing non-core-competencies means anything that's a commodity — services which are not considered a differentiator in enterprise strategies. While this sounds plausible on the surface, it's not wise.

39. Meyer, N. Dean. *Outsourcing: how to make vendors work for your shareholders.* Danbury, CT: NDMA Publishing. 1999.

40. For more on decentralization, see: Meyer, N. Dean. *Decentralization: fantasies, failings, and fundamentals.* Danbury, CT: NDMA Publishing. 1998.

41. Meyer, N. Dean. *Fast Track to Changing Corporate Culture.* Danbury, CT: NDMA Publishing. 2003.

42. In addition to healthy structure, a shared-services provider must give clients control over what they buy through a properly designed internal economy. See: Meyer, N. Dean. *Internal Market Economics: practical resource-governance processes based on principles we all believe in.* Danbury, CT: NDMA Publishing. 2013.

43. Sayles, Leonard R. "Matrix Organization: The Structure with a Future." *Organizational Dynamics.* Autumn, 1976. Pages 2-17.

44. I'm grateful to Rich Lack for this case study.

45. Peter Drucker, an expert on management (not organizational engineering), inappropriately recommends a structural separation between thinking and doing, new and old products. See: Drucker, Peter F. *Innovation and Entrepreneurship.* New York: Harper & Row. 1985. Pages 161-170.

46. "Alfred Sloan, Move Over." *Chief Executive.* Number 87, July/August, 1993.

47. Hirschhorn, Larry and Thomas Gilmore. "The New Boundaries of the 'Boundaryless' Company." *Harvard Business Review.* May-June, 1992. Pages 104-115.

48. Burns, Tom and George M. Stalker. *The Management of Innovation.* London: Tavistock. 1961.

49. Earlier manifestations include:
"Holographic" organization: Mackenzie, Kenneth D. *The Organizational Hologram: the effective management of organizational change.* Boston: Kluwer. 1991.
"Holonic" organization: Matthews, J. A. "Holonic Organisational Architectures." *Human Systems Management.* Number 15, 1996. Pages 1-29.

50. Satell, Greg. "What Makes an Organization 'Networked'?" *Harvard Business Review.* June 8, 2015.

51. *Valve Handbook for New Employees.* Bellevue, WA: Valve Corporation. 2012.

52. Warr, Philippa. "Former Valve Employee: 'It Felt a Lot Like High School.'" *Wired.* July 9, 2013.

53. Mittelman, Melissa. "Why GitHub Finally Abandoned Its Bossless Workplace." *Bloomberg Technology.* September 6, 2016.

54. Miles, Raymond E., Grant Miles, and Charles C. Snow. *Collaborative Entrepreneurship.* Palo Alto, CA: Stanford Business Books. 2005.

55. Bureaucracy was originally defined by: Weber, Max. *Economy and Society.* 1922. Edited by Guenther Roth and Claus Wittich; and republished by University of California Press. 1968.
    i. Fixed and official jurisdictions (domains); official duties, rules of authority; qualifications for positions
    ii. The chain of command, i.e., a hierarchy of supervision authority; the right to appeal a decisions to higher authorities
    iii. Management is based on written documentation
    iv. Thorough and expert training
    v. Job duties are everyone's primary focus
    vi. Stable rules
    This is not a description of disempowering management, excessive rules, and unnecessary paperwork, as people use the term today.

56. Mintzberg called this "adhocracy": Mintzberg, Henry. *The Structuring of Organizations.* Englewood Cliffs, NJ: Prentice-Hall. 1979. Chapter 21.

57. Badal, Jaclyne. "Can a Company Be Run as a Democracy?" *Wall Street Journal.* April 23, 2007.
    Robertson, Brian. "Evolving Organization." *Integral Leadership Review.* Volume 7, number 3, June, 2007.
    The "Constitution" is now published by HolacracyOne LLC.

ENDNOTES (References)

58. Compte, Auguste. *A General View of Positivism.* Trubner and Co. 1865. (Reissued by Cambridge University Press. 2009.) And: Ward, Lester Frank. *Penn Monthly.* 1881. Also: Ward, Lester Frank. *Dynamic Sociology.* New York: D. Appleton & Co. 1883.

59. Boeke, Cornelis (Kees). "Sociocracy: Democracy as It Might Be." 1945. And: Endenburg, Gerard. *Sociocracy: the organization of decision-making.* Delft, The Netherlands: Eburon. 1998.

60. Jaques, Elliot. *Requisite Organization.* Arlington, VA: Cason Hall & Co. 1989.

61. HolacracyOne. www.Holacracy.org/how-it-works

62. Gelles, David. "At Zappos, Pushing Shoes and a Vision." *New York Times.* July 17, 2015.

63. I'm grateful to Lloyd Nimetz for this case study.

64. Feloni, Richard. "How Zappos Decides How Much to Pay Employees Under its New 'Self-management' System." *Business Insider.* July 24, 2015.

65. Gillette, Halbert P. *Handbook of Cost Data for Contractors and Engineers: A Reference Book Giving Methods of Construction and Actual Costs of Materials and Labor on Numerous Engineering Works.* New York, NY: M. C. Clark. 1905.

66. Pascale, Richard. *Managing on the Edge.* Touchstone Books. 1991.

67. Meyer, N. Dean. *Internal Market Economics: practical resource-governance processes based on principles we all believe in.* Danbury, CT: NDMA Publishing. 2013.

68. Dunbar, Robin. *How Many Friends Does One Person Need?: Dunbar's number and other evolutionary quirks.* London: Faber and Faber. 2010. Also: Hill, Russell A. and Robin Dunbar. "Social Network Size In Humans." *Human Nature.* Number 14:53, 2003.

69. Dunbar, Robin. "Robin Dunbar on Dunbar Numbers." Interview on *Social Science Bites.* November 4, 2013.

70. "Post-it" is a trademark of 3M Corporation.

71. Zhou, Wei-Xing, Didier Sornette, Russell A. Hill, and Robin Dunbar. "Discrete hierarchical organization of social group sizes." *Proceedings of the Royal Society B: Biological Sciences.* Volume 272(1561), February 17, 2005. Pages 439-444.

72. In the Human Resources literature, there's much controversy about how to calculate span of control, and what the right numbers are. See: Mintzberg, Henry. *The Structuring of Organizations.* Englewood Cliffs, NJ: Prentice-Hall. 1979. Chapter 8.

    Viewing it through the lens of cybernetics, as a matter of requisite variety, simplifies the concept. Any factor which increases the variety a group has to process (complexity, volume, time pressures) reduces the appropriate span of control.

73. A deputy should be an assistant to the executive, not a supervisor of the next tier of leadership (which would leave the executive with a span of one, and distance the executive from his/her leaders).

74. In IT, for example, layers of Asset-based Service Providers include the data center and operator services, computing and storage services, middleware and database services, applications hosting, subscription-based services such as email and analytics tools, and access to shared data (business intelligence).

75. I'm grateful to Sergio Paiz, CEO of PDC, for this case study and its associated insights.

76. See, for example: Crosby, Philip B. *Quality Is Free: The Art of Making Quality Certain.* New York, NY: McGraw-Hill. 1979.

77. Mintzberg, Henry. *Structure in Fives: designing effective organizations.* London: Pearson. 1983. Page 2.

78. Mintzberg, Henry. *Structure in Fives: designing effective organizations.* London: Pearson. 1983. Page 4.

79. I'm grateful to Marcy Reman for this case study.

80. Wheatley, Margaret J. *Leadership and the New Science.* San Francisco, CA: Berrett-Koehler Publishers. 2003.

81. Hammer, Michael and James Champy. *Reengineering the Corporation: a manifesto for business revolution.* New York, NY: HarperBusiness Essentials. 1993.

ENDNOTES (References)

82. Taylor, James C. and David F. Felten. *Performance by Design: Sociotechnical Systems in North America.* Upper Saddle River, NJ: Prentice-Hall. 1993.

83. Hamel, Gary and C.K. Prahalad. *Competing for the Future.* Cambridge, MA: Harvard Business Review Press. 1996.

84. External vendors, contract employees, and consultants are not considered subcontractors in this context; rather, they're treated as part of a group's staff; they live within that group's domain; and they're managed by that group. For more on this, see: Meyer, N. Dean. *Outsourcing: how to make vendors work for your shareholders.* Danbury, CT: NDMA Publishing. 1999.

85. Note that the terms "customer" and "supplier" refer to a relationship within the context of a specific deliverable. Group A may be a customer to Group B on one project, and a supplier to Group B on another.

    Miles and Snow called this the "market-matrix" organization: Miles, Raymond E. and Charles C. Snow. *Organizational Strategy, Structure, and Process.* New York, NY: McGraw-Hill. 1978.

    The resource-governance implications and mechanisms that make this work are described in: Meyer, N. Dean. *Internal Market Economics: practical resource-governance processes based on principles we all believe in.* Danbury, CT: NDMA Publishing. 2013.

86. Stevenson, Richard W. "Automotive Aristocrat Trims Down." *New York Times.* March 8, 1994.

87. Meyer, N. Dean. *Internal Market Economics: practical resource-governance processes based on principles we all believe in.* Danbury, CT: NDMA Publishing. 2013. Chapter 12.

88. Meyer, N. Dean. *Fast Track to Changing Corporate Culture.* Danbury, CT: NDMA Publishing. 2003.

89. One popular change-management model is "ADKAR" proposed by Prosci Inc. It's focused on *how to sell a change decided by executives,* rather than a participative process of change such as is described here.
    - **A**wareness of the business reasons for change. Awareness is the goal/outcome of early communications related to an organizational change.
    - **D**esire to engage and participate in the change. Desire is the goal/outcome of sponsorship and resistance management.
    - **K**nowledge about how to change. Knowledge is the goal/outcome of training and coaching.
    - **A**bility to realize or implement the change at the required performance level. Ability is the goal/outcome of additional coaching, practice and time.
    - **R**einforcement to ensure change sticks. Reinforcement is the goal/outcome of adoption measurement, corrective action and recognition of successful change.

90. Senge, Peter M. *The Fifth Discipline: The Art & Practice of the Learning Organization.* New York, NY: Doubleday. 1990. Page 155.

91. Bridges, William. *Managing Transitions: making the most of change.* Boston, MA: Nicholas Brealey. 2010.

92. Chip and Dan Heath suggest six attributes of communications that make a new idea stick, all of which can be incorporated into the change process: Heath, Chip and Dan Heath. *Made to Stick: Why Some Ideas Survive and Others Die.* Random House. 2007.
    - **Simple:** Simplicity isn't about dumbing down; it's about prioritizing. What's the core of your message?

      At its most basic, a principle-based structure says, "Everybody is an entrepreneur."
    - **Unexpected:** To get attention, violate a schema. To hold attention, use curiosity gaps. Before your message can stick, your audience has to want it.

      An example is, "We don't pay you to do your job; we pay you for results. So what do you "sell"?"
    - **Concrete:** To be concrete, use sensory language. (Think Aesop's fables.) Paint a mental picture.

      An example is this analogy: "The general contractor who builds a house hires the plumber and electrician. These subcontractors are accountable for their own deliverables. They choose their supply stores, who in turn are accountable for delivering needed parts."

- **Credible:** Ideas can get credibility from outside (authorities or anti-authorities) or from within, using human-scale statistics or vivid details. Let people "try before they buy."

   The Rainbow Exercise crystalizes what's not working in the current structure.

- **Emotional:** People care about people, not numbers. Don't forget the WIIFY (What's In It For You).

   The reason for restructuring is simple: "This new organization is designed to be the employer of choice to our staff, and supplier of choice to our customers."

- **Stories:** Stories drive action through simulation (what to do) and inspiration (the motivation to do it). Springboard stories help people see how an existing problem might change.

   Walk-throughs during the planning process are such simulations, and show how the new design will benefit real-life projects.

93. In fact, beyond conventional 360-degree reviews which only gather generic input from peers, in a principle-based structure, you can ask a group's customers and suppliers different questions. Customer satisfaction is different from suppliers' assessment of team-leadership skills.

94. The documentation, part of *The Principle-based Structure Knowledge Library,* is available under license. Contact <ndma@ndma.com> for more information.

95. Posting all the jobs at this point in the process is not appropriate. Anyone selected from outside the organization could displace a current employee. They wouldn't have been part of the design process, and hence wouldn't have the depth of understanding and commitment as do the candidates who participated. And an open posting would cause a significant delay at a critical point in the process.

   If open headcount permits, you may post for participation in the candidate pool before the process starts. Or you may post any remaining open positions after leaders have been selected from the candidate pool, where the competencies aren't found within the organization.

96. Bridges, William. *Managing Transitions: making the most of change.* Boston, MA: Nicholas Brealey. 2010.

97. De Pree, Max. *Leadership is an Art.* Crown Business. 2004.

98. Marsh, Steve. "A Car for Every Driver." *Delta Sky.* January, 2017. Page 107.

99. I'm grateful to Marcy Reman for this case study.

100. McCormack, Ade G. *Beyond Nine to Five.* Buckinghamshire, UK: Auridian Press. 2015.

101. Meyer, N. Dean. *RoadMap: how to understand, diagnose, and fix your organization.* Danbury, CT: NDMA Publishing. 1998.

102. Meyer, N. Dean. *Fast Track to Changing Corporate Culture.* Danbury, CT: NDMA Publishing. 2003.

103. Meyer, N. Dean. *Internal Market Economics: practical resource-governance processes based on principles we all believe in.* Danbury, CT: NDMA Publishing. 2013.

104. Hummel, Charles E. *Tyranny of the Urgent.* Downers Grove, IL: InterVarsity Press. 1967.

105. Beer, Stafford. *The Heart of the Enterprise.* Chichester, UK: John Wiley & Sons. 1979. Page 7.

106. The recognition that there's more to economic decision-making than money dates back to Adam Smith's 1759 book, *The Theory of Moral Sentiments*, in which he noted that people are rewarded by pride in their own morality, such as the good feeling that results from serving others.

    More recently, the field of behavior economics explores the many reasons people deviate from purely rational economic decision-making.

    Daniel Kahneman and Amos Tversky explored subjective perceptions of risk in their paper: Kahneman, Daniel and Amos Tversky. "Prospect Theory: An Analysis of Decision Under Risk" *Econometrica.* Number 47(2), March, 1979. Pages 263-291.

    Subsequently, Kahneman wrote a best-selling book that summarizes his research on why decision-making is not purely rational: Kahneman, Daniel. *Thinking, Fast and Slow.* Macmillan. 2011.

    Beyond Kahneman's work on distortions in decision making, Dan Ariely reviewed the many reasons people deviate from purely economic trade-offs in his book: Ariely, Dan. *Predictably Irrational.* Harper Collins. 2008.

107. Smith, Adam. *The Wealth of Nations.* 1776.

108. Ashby, W. Ross. *Introduction to Cybernetics.* London: Chapman & Hall. 1956.

INDEX

360-degree reviews *323*
accountability, and authority *23*
acquisitions *342*
adhocracy *211 (endnote)*
ADKAR *317 (endnote), 402*
administration, internal *112*
agile development group *198*
airline case study *9*
alignment
    priorities of internal service provider with business *67*
    with clients' strategies *126, 127*
altruism *8, 388*
Anderson, Dave *326*
Applications Engineers *104, 105*
applications, in IT *106*
    rationalization of *34*
architecture, in IT *45, 122*
Ariely, Dan *387 (endnote)*
Aristotle *315*
Ashby, W. Ross *37 (endnote), 389*
assessment, as distinct from Audit *119, 136*
asset management *383*
Asset-based Service Providers *109*
assignments
    of people into new structure *330, 333*
    to two jobs *248*
Audit *136*
    as distinct from assessment *119, 136*
    as distinct from testing *136*
    competencies of *139*
    place in structure *140*
    scope *138*
    versus service *75*
Audit Response Coordinator *118*
authority, and accountability *23*
Babington, Thomas *45*
Badal, Jaclyne *213 (endnote)*
Barra, Barra *358*

Base Engineers *104, 105*
    and Coordinators *73*
basis for substructure *60, 63*
    lines of business *84*
Beer, Stafford *385*
behavioral economics *387, 404*
Benci, John *329*
benefits measurement, methods *133*
benefits of healthy structure *2, 352*
Bergeron, Peter *71 (endnote)*
best practices *294*
bi-modal IT *198*
Boeke, Cornelis (Kees) *213*
Bohr, Niels *38*
bonuses *323*
boundaries *see domains*
boundary, between Sales and Engineering *341*
boundaryless organization *206*
brand managers *302, 311*
Brandeau, Greg *1 (endnote)*
Bridges, William *343 (endnote)*
broker *381*
    of vendor services *162*
Brown, Gordon Stanley *14*
Brown, Ken *136 (endnote)*
budget
    CFO deciding is disempowering *25*
    planning *382*
bureaucracy, definition *398*
Burns, Tom *43 (endnote), 207 (endnote)*
business analysts *130*
business continuity *270, 276, 383*
Business Continuity Coordinator *119*
business diagnosis
    methods *133*
    role of Sales and Marketing *127*
business process
    as a basis for structure, case study *60*
    facilitation versus change *358*
    reengineering (BPR) *294*
business relationship managers *see Sales*

INDEX 407

business within a business   *84*
    and teamwork   *301*
    culture of   *357*
capacity management   *94, 383*
career counseling   *383*
career paths   *81*
    problems with programmer pools   *205*
Carroll, Lewis   *22*
catalogs
    benefits of   *308*
    place in implementation process   *331, 333*
cellular organization   *207*
centers of excellence, to ameliorate the problems of decentralization   *177*
centralization   *see decentralization*
Cezanne, Paul   *141*
Champy, James   *294 (endnote)*
Chandler, Alfred D.   *154*
change management   *317*
    climate for change   *318*
    guarantees to staff   *320*
    making changes stick   *322*
    transitions (the human reactions to change)   *321*
Chief Digital Officer   *127, 131, 396*
Chief Information Security Officer (CISO)   *120*
Clean Sheet approach   *224*
client, definition   *379*
climate for change   *see change management*
cloud broker   *162*
coaching   *383*
    without disempowering   *29*
coffee, story of project management   *300*
communications, in structural change   *322*
compensation, loss of   *321*
competencies
    Audit   *139*
    Coordinators   *124*
    Engineers   *106*
    Sales and Marketing   *133*
    Service Providers   *112*
competition, internal   *46*
    sources of   *87*
compliance group   *270*
component solutions   *72*

Compte, Auguste *213*
conflicts of interests *67*
    consequences *75*
    types of *69*
Confucius *153*
consolidations *342*
Consultancy *see Sales*
contingency theories *13, 393*
continual improvement, impacts of structure *357*
contractors, versus subcontractors, role in organization *302 (endnote)*
contracts, internal *382*
control-oriented functions *270*
Conway, Melvin E. (Conway's Law) *83 (endnote)*
coordination mechanisms, alternatives *290*
Coordinators *115*
    and Base Engineers *73*
    business-oriented *116*
    competencies of *124*
    function-specific *123*
    Organizational Effectiveness *239*
    technical *121*
    types of *116*
cost *3, 4, 25, 27, 40, 47, 61, 61, 65, 81, 88, 94, 147, 148, 162, 171, 171, 172, 174, 180, 194, 209, 210, 220, 352, 357, see also productivity*
Coughran, Bill *1*
Crosby, Philip B. *276 (endnote)*
culture *365*
    impacts of structure *356*
    of teamwork *313*
customer focus *3, 4, 62, 88, 91, 93, 148, 165, 184, 356*
customer service *111*
    accountability for resolution of problems *25*
customer, definition *379*
customer-centric structure *32*
cyber-security *276*
cybernetics *385*
Cyert, Richard *156 (endnote)*
De Pree, Max *351*

INDEX

decentralization *171*
    dotted lines to shared-services staff *175*
    encouraged by structure *35*
    engineering, case study *173*
    federated model *175*
    IT, case study *174*
    manufacturing, case study *173*
    pressures for *32*
    treated as People-based Service Providers *112*
decision making, within groups *383*
demand management *67, 229*
Design Patterns Coordinator *121*
design, versus repair *103*
development (product), versus maintenance *103*
development (training) planning *383*
development center, in IT *198*
DevOps *283*
digital enterprise *127, 131, 390*
discipline, of employees *384*
domains *45*
    gaps, consequences *50*
    overlaps *46*
    place in implementation process *330, 333*
    within groups *383*
Dong, Walter *27 (endnote)*
dotted lines *184*
    decentralized staff to shared-services leader *175*
    shared-services staff to business leaders *185*
Drucker, Peter F. *88 (endnote), 154 (endnote), 194 (endnote)*
du jour organization *15*
dual reporting *184, see also dotted lines*
    distinct from temporary duty (shared people) *260*
Dugan, Ianthe Jeanne *136 (endnote)*
Dunbar, Robin *230, 231 (endnote), 244 (endnote)*
effectiveness, impact of structure *354*
efficiency *88*
    impact of structure *352*
Einstein, Albert *369*
emerging technologies group *189*
Emerson, Ralph Waldo *21*
employer of choice *359*
Employment Policy Coordinator *123*

empowerment *23, 26*
    role of management *29*
Endenburg, Gerard *213*
Engineers *103*
    and Service Providers, combining *69*
    Applications *105*
    Base *105*
        gap *48*
        versus Applications *72*
    competencies of *106*
    decentralization, case study *173*
    services of *103*
    versus Sales *74*
enlightened self interests *388*
enterprise architecture, in IT *45, 121, 122*
enterprise, definition *377*
entrepreneurship *85*
    benefits of structure *357*
    traits of *88*
execution separate from policy *23 (endnote), 393*
extended staffing *88, 162*
federated model *175*
feedback loops *386*
feedback, from customers *382*
Feloni, Richard *219 (endnote)*
Felten, David F. *294 (endnote)*
field technicians *264*
financial management *382*
flat structure *247*
focus, lack of *51*
Ford, Henry *32*
Fouraker, Lawrence E. *43 (endnote)*
Franklin, Benjamin *353*
Frost, Robert *51*
Galbraith, Jay R. *158*
gaps
    case study *48*
    consequences *50*
Garfield, James A. *279*
Gelb, Steve *9 (endnote)*
Gelles, David *218 (endnote)*
General Electric *206*
generalist *32, 38*

geography, dispersion *262*
Gillette, Halbert P. *223*
Gilmore, Thomas *206 (endnote)*
GitHub *207*
Golden Rule *23*
governance *270*
    group *270*
        case study *67*
    to ameliorate the problems of decentralization *177*
ground-rules *320*
group, definition *378*
guarantees to staff during restructuring *320*
Guest, David *38 (endnote)*
Hamel, Gary *296*
Hammer, Michael *294 (endnote)*
hand-offs *304*
    minimizing *43, 78*
headcount
    impact of structure *352*
    load balancing *259*
    reductions *321*
    sharing of *see temporary assignments*
Heath, Chip and Dan *322 (endnote)*
help desk *see customer service*
hierarchical, versus organic *211*
Hill, Linda *1 (endnote)*
Hill, Russell A. *230 (endnote), 244 (endnote)*
hiring *383*
Hirschhorn, Larry *206 (endnote)*
Holacracy *213*
Houghton, William T. *312*
Hummel, Charles E. *373*
implementation process *325*
    selection of people *330, 333*
informal organization *7*
Information Policy Coordinator *123*
infrastructure *93*
    in IT *106*
    owners of *109*

innovation *3, 4, 9, 40, 47, 47, 50, 65, 80, 88, 94, 146, 147, 148, 150, 172, 174, 181, 189, 193, 194, 199, 200, 209, 210, 216, 218, 355, 356, 357*
    and specialization *40*
    combining with operations *9, 69*
    group *189*
    intentional damper on *113*
    invention *71*
    lack of, root causes *229*
    problems with *9*
    separate group *189*
inspection *276*
    group, case study *84*
intake, of work, not a group in structure *67*
integration, of suppliers *162*
internal economy *367*
    how it works *228*
    problems when addressed through structure *67, 189, 196, 201, 228, 229*
internal market economics *302 (endnote)*
internal service provider, definition *377*
intrapreneurship *88*
invention *9, 71, see also innovation*
investment-based budgeting *117 (endnote)*
IT, decentralization, case study *174*
ITIL *283, 294*
    as a basis for structure, case study *78*
Jacka, Mike *54 (endnote)*
Jaques, Elliot *213*
job narrowing *41*
Johnson, Samuel *39*
Kahneman, Daniel *387 (endnote)*
Keller, Paulette *54 (endnote)*
KPIs *368*
Lack, Rich *193 (endnote)*
leaderless organization *213*
Lean *283, 294*
learning, separate from doing *189*
Letzkus, Brian *33 (endnote)*
Lindblom, Ed *297, 301*
Lineback, Kent *1 (endnote)*
linkage, organization's products to clients' business *133*
load balancing *81, 259*
long-term versus short-term projects *198*
Macaulay, Baron *45*

Mackenzie, Kenneth D. *207 (endnote)*
maintenance
    separation from development *193*
    versus development *103*
management, and empowerment *29*
managing staff *383*
manufacturing, decentralization, case study *173*
March, James G. *156 (endnote)*
Marketing (Building Block) *126, 132*
marketing
    as a management task *382*
    promotions, coordination across product lines *122*
matrix structure *186*
Matthews, J. A. *207 (endnote)*
McCormack, Ade *360*
means versus ends, as a basis for structure *198*
mentoring *80*
mergers *342*
meta-process *300*
methods
    as a basis for structure *198*
    management task *383*
    organizational system *368*
    versus structure *201*
metrics *368*
    and structure *51*
    shared *257, 388*
Miles, Grant *208 (endnote)*
Miles, Raymond E. *2 (endnote), 208 (endnote), 302 (endnote)*
millennials *360*
Mintzberg, Henry *31 (endnote), 41 (endnote), 63 (endnote), 211 (endnote), 245 (endnote), 279, 290*
Mittelman, Melissa *207 (endnote)*
monopoly, for internal service providers *87*
morale *see motivation and morale*
motivation and morale *3, 4, 25, 26, 40, 41, 47, 51, 55, 76, 81, 147, 148, 150, 186, 190, 195, 195, 200, 204, 209, 210, 219, 348, 355, 359*
    and specialization *40*
mutual adjustment *287, 290, 297, 300*
needs assessment, methods *133*
network organization *207*
Nimetz, Lloyd *218 (endnote)*
objective setting *383*

Operational Planning Coordinator   *117*
operations, combining with innovation   *9, 69*
organic organization   *207*
    versus hierarchical   *211*
Organizational Effectiveness Coordinator   *118, 239*
organizations   *376*
    as systems   *385*
    du jour   *15*
    five systems of   *363*
    informal   *7*
    why they exist   *32, 39*
outsourcing
    how to manage vendors   *162*
    problems with   *158*
    strategic   *88*
    supplier integration   *162*
    versus subcontractors   *302 (endnote)*
    when it's appropriate   *161*
overlaps   *46*
Paiz, Sergio   *xiii, 1, 73, 251 (endnote), 306, 362, 362*
participation, in structural change, value of   *319*
partnership, between internal service providers and clients   *89, 91*
Pascale, Richard   *225*
passive order takers   *92, 128, 130*
PDC   *xiii, 73, 362*
People-based Service Providers   *110*
performance appraisals   *383*
    self-managed groups   *257*
    temporary duty (shared people)   *260*
performance management   *29, 51, 218, 323, 383*
Pinchot, Gifford   *88 (endnote)*
plan-build-run   *189*
planning   *115*
    business strategy   *382*
    group   *189*
Plato   *44*
policy separate from execution   *23 (endnote), 393*
posting jobs   *330 (endnote), 403*
Prahalad, C.K.   *296*
Preston-Werner, Tom   *207*
pricing, coordination across product lines   *122*
prime contractor   *301*

INDEX

priority setting *229*
    giving clients control *228*
proactive entrepreneurship *92*
problems, structural *142*
process
    business, as a basis for structure, case study *60*
    internal, as a basis for structure, case study *78*
process engineering *60, 294*
    no need for *305*
process owners *23*
    no need for *304*
product design *103*
product management *94, 382*
productivity *40, 47, 50, 80, 146, 147, 150, 169, 199, 218, 353, 355*
professional practices *383*
professional synergies *78*
project management
    standardized process is disempowering *28*
    walk-throughs as first step *309*
project managers
    no need for super project managers *305*
    place in structure *239*
    role of *52, 57, 111*
    who run projects *267*
project plan, via walk-throughs *309*
project team *see team*
proposal writing *382*
purpose-specific products *72, 105*
quality *3, 4, 40, 47, 50, 80, 88, 94, 146, 147, 148, 150, 169, 174, 181, 194, 199, 200, 204, 209, 218, 270, 355, 356*
    and specialization *40*
    assurance group *276*
quick versus slow groups *198*
RACI *54*
rainbow
    in remote locations *262*
    people *225, 259*
Rainbow Analysis *142*
rapid applications development group *198*
recommendations, without sacrificing customer focus *93*
reengineering, business process *see process engineering*
regulatory compliance *270*
Regulatory Compliance Coordinator *119*

reinforcement, in structural change *322*
relationship managers *see Sales*
relationships, with clients *126, 127*
Reman, Marcy *291 (endnote), 358 (endnote)*
remote locations *262*
reorganizations, repeated *15*
repair
    separation from development *193*
    versus design *103*
reporting, to management *383*
requirements planning *130*
Research Coordinator *117*
research group *189*
resource management, management task *384*
resource-governance processes *see internal economy*
restructuring, implementation process *325*
results orientation *27, 55, 88*
reward systems *323, 368, 383*
risk *40, 47, 50, 88, 146, 147, 148, 150*
Robertson, Brian *213*
Roig, Randy *84 (endnote)*
roles and responsibilities, problems with *52*
Rolls-Royce *306*
safety *270*
    group, case study *84*
sales *382*
Sales and Marketing *126*
    Account *129*
    competencies of *133*
    done by senior managers *165*
    Function *131*
    Marketing *132*
    Retail *130*
    substructured by product line *64*
    versus Engineers *74*
    why they fail *336*
sales incentives, coordination across product lines *122*
same name in two boxes *248*
Satell, Greg *207 (endnote)*
Sayles, Leonard R. *187 (endnote)*
scalable *355*
scapegoat trap *276*
scattered campuses, the antidote *79*

INDEX

science *13*
    of structure *13, 391*
        importance of *14*
second-class citizens *99, 191, 204*
security *270, 276*
Security Coordinator *120*
selection of people into new structure *330, 333*
self-directed group *379*
self-forming teams *288*
self-managed group *256, 379*
self-managed team *288, 379*
sell *126, 381*
    consultative selling *127*
    partnership selling *127*
    strategic selling *127*
Senge, Peter *319*
service broker *162*
service management *112*
Service Providers *109*
    and Engineers, combining *69*
    Asset-based *109*
    competencies of *112*
    People-based *110*
service, versus Audit *75*
shadow IT *32*
shared services *88, 171, 180, 342*
    benefits *180*
    buying from other departments *237*
    role of *177*
short-term versus long-term projects *198*
silo organization *43, 291*
    direct supervision to coordinate work *292*
    Dunbar's number *230*
    manufacturing *291*
    process-centric groups *78*
    strategy-centric groups *155*
Simon, Herbert *156 (endnote)*
Simons, Preston T. *17, 179*
site coordinator *263, 384*
Six Sigma *283, 294*
Smith, Adam *39, 387 (endnote), 388 (endnote)*
Snow, Charles C. *2 (endnote), 208 (endnote), 302 (endnote)*
Sociocracy *213*

Socrates *97*
Sornette, Didier *244 (endnote)*
sourcing, strategic *88*
span of control *244*
    appropriate *246*
    calculating *245*
    target levels *246*
specialist, definition *37*
specialization *32*
    benefits of *39*
    concerns about *41*
    dependence on teamwork *42*
    relationship to structure *63*
speed *3, 4, 40, 47, 50, 61, 80, 146, 147, 148, 150, 174, 198, 199, 204*
Stalker, George M. *43 (endnote), 207 (endnote)*
standardization, as coordinating mechanism *290*
Standards Coordinator *121*
Stevenson, Richard W. *306 (endnote)*
Stevenson, Robert Louis *302*
stick, making changes stick *322*
Stopford, John M. *43 (endnote)*
strategic alignment *126, 127*
strategic planning *115*
strategic sourcing *88*
strategic value *3, 4, 62, 92, 154, 158, 167, 172, 210*
    role of Sales and Marketing *127*
Strategy Planning Coordinator *116*
strategy, as a basis for structure *154*
structure
    benefits *352*
    by business process, case study *60*
    by client organization *32*
    by internal process, case study *78*
    by strategy *154*
    by tasks *55*
    definition *279, 378*
    diagnosing problems *142*
    frequent reorganizations *15*
    implementation process *325*
    importance of *9*
    matrix *186*
    new versus old *193*
    plan-build-run *189*

symptoms of problems with *4*
    versus methods *201*
    within groups *382*
subcontractor *302, 383*
substructure, basis for *60, 63, 149*
super project managers, no need for *305*
supervision, as a coordinating mechanism *290, 291*
supplier integration *162*
support functions see internal service provider
symptoms of structural problems *4, 142*
synergies
    enterprise, from support functions *172*
    professional *78, 80*
    types of *79*
system, definition *385*
systems science see cybernetics
T-shaped specialist *38*
tasks, jobs designed around *55*
Taylor, James C. *294 (endnote)*
team, definition *378*
teamwork *3, 4, 32, 48, 58, 88, 191, 195, 210, 220, 355*
    and specialization *42*
    as a substitute for clear structure *206*
    concerns about *43*
    culture of *313*
    high-performance *287, 300*
    incentive for, no need *323*
    self-forming teams *288*
    through internal customer-supplier relationships *301*
    through mutual adjustment *287, 297, 300*
    under common boss *43*
technical excellence, versus unbiased diagnosis *74*
technicians, field *111, 264*
Technologists see Engineers
temporary assignments see temporary duty
temporary duty (shared people) *259*
    distinct from dual reporting *260*
    in remote locations *265*
testing *276*
    as distinct from Audit *136*
Thomas, Noel *325*
tier, of leadership *246*
tools *368*

Total Quality Management *276, 356*
traffic circles, New Jersey *285*
training *111*
transformation
    five organizational systems of *363*
    impacts of structure *356*
    planning *363*
transitions (the human reactions to change) *321, 384*
    place in implementation process *330, 333*
Treacy, Michael *158*
Truelove, Emily *1 (endnote)*
Tversky, Amos *387 (endnote)*
Tweaks approach *224*
Tyranny of the Urgent *373*
unbiased diagnosis, versus technical excellence *74*
Urwick, Lyndall *25 (endnote)*
Valve Corporation *207*
variety (from cybernetics) *36, 389*
vendors, versus subcontractors, role in organization *302 (endnote)*
walk-throughs *309*
    place in implementation process *331, 333*
Wanstrath, Chris *207*
Ward, Lester Frank *213*
Warr, Philippa *207 (endnote)*
Weber, Max *211 (endnote)*
Welch, Jack *1, 206*
Wheatley, Margaret J. *292*
Wiersema, Fred *158*
Wilson, Edward O. *320*
work allocation *384*
work intake, not a group in structure *67*
working manager *245*
writers *111*
Zappos *213*
Zhou, Wei-Xing *244 (endnote)*

Other Books by N. Dean Meyer

Internal Market Economics: practical resource-governance processes based on principles we all believe in. 2013. (book)

Full-cost Maturity Model. 2007, 2008. (report)

An Introduction to the Business-Within-a-Business Paradigm. 2002. (monograph)

RoadMap: how to understand, diagnose, and fix your organization. 1998. (book)

Fast Track to Changing Corporate Culture. 2003. (monograph)

Meyer's Rules of Order: how to hold highly productive business meetings. 2001. (pocket book)

Outsourcing: how to make vendors work for your shareholders. 1999. (book)

Decentralization: fantasies, failings, and fundamentals. 1998. (book)

Downsizing Without Destroying: how to trim what your organization does rather than destroy its ability to do anything at all. 2003, 2008. (monograph)

The New Lexicon of Leadership: dictionary of terms used in leadership and organizational design. 2012. (report)

The Information Edge with Mary E. Boone. 1987. (book)

About the Author

Dean Meyer is one of the original proponents of the business-within-a-business paradigm, where every managerial group is an entrepreneurship funded to produce products and services for customers. He has implemented this philosophy in corporate, government, and non-profit organizations through the careful design of culture, organizational structure, and resource-governance processes.

Dean is the author of eight books (six still in print), numerous monographs, and countless articles.

His book, *Internal Market Economics*, pioneered the application of market economics inside companies to design non-bureaucratic resource-governance processes. And he invented *FullCost,* a business planning method and tool that produces a budget for what an organization proposes to sell (as well as what it proposes to spend).

He researched and applied an entirely new science of organizational structure, as documented in this book.

And he developed an approach to corporate culture that leads to meaningful change in less than a year.

Dean coaches executives on organizational issues, and facilitates transformation processes. Both a visionary and a mechanic, he paints a clear picture of how organizations should work, and implements that vision through pragmatic, participative change processes.

Dean is a native of San Francisco and resident of Connecticut. He received a BS from the University of California at Berkeley, and a MBA from Stanford University.